SpringerBriefs in Philosophy

SpringerBriefs present concise summaries of cutting-edge research and practical applications across a wide spectrum of fields. Featuring compact volumes of 50 to 125 pages, the series covers a range of content from professional to academic. Typical topics might include:

- A timely report of state-of-the art analytical techniques
- A bridge between new research results, as published in journal articles, and a contextual literature review
- A snapshot of a hot or emerging topic
- An in-depth case study or clinical example
- A presentation of core concepts that students must understand in order to make independent contributions

SpringerBriefs in Philosophy cover a broad range of philosophical fields including: Philosophy of Science, Logic, Non-Western Thinking and Western Philosophy. We also consider biographies, full or partial, of key thinkers and pioneers.

SpringerBriefs are characterized by fast, global electronic dissemination, standard publishing contracts, standardized manuscript preparation and formatting guidelines, and expedited production schedules. Both solicited and unsolicited manuscripts are considered for publication in the SpringerBriefs in Philosophy series. Potential authors are warmly invited to complete and submit the Briefs Author Proposal form. All projects will be submitted to editorial review by external advisors.

SpringerBriefs are characterized by expedited production schedules with the aim for publication 8 to 12 weeks after acceptance and fast, global electronic dissemination through our online platform SpringerLink. The standard concise author contracts guarantee that

- an individual ISBN is assigned to each manuscript
- each manuscript is copyrighted in the name of the author
- the author retains the right to post the pre-publication version on his/her website or that of his/her institution.

Andrzej Zaporowski

Crack and Culture: On Representations of Movement in Anthropology and Philosophy

Andrzej Zaporowski
Adam Mickiewicz University
Poznan, Poland

ISSN 2211-4548　　　　　　　ISSN 2211-4556　(electronic)
SpringerBriefs in Philosophy
ISBN 978-3-031-83421-9　　　ISBN 978-3-031-83422-6　(eBook)
https://doi.org/10.1007/978-3-031-83422-6

© The Editor(s) (if applicable) and The Author(s), under exclusive license to Springer Nature Switzerland AG 2025

This work is subject to copyright. All rights are solely and exclusively licensed by the Publisher, whether the whole or part of the material is concerned, specifically the rights of translation, reprinting, reuse of illustrations, recitation, broadcasting, reproduction on microfilms or in any other physical way, and transmission or information storage and retrieval, electronic adaptation, computer software, or by similar or dissimilar methodology now known or hereafter developed.
The use of general descriptive names, registered names, trademarks, service marks, etc. in this publication does not imply, even in the absence of a specific statement, that such names are exempt from the relevant protective laws and regulations and therefore free for general use.
The publisher, the authors and the editors are safe to assume that the advice and information in this book are believed to be true and accurate at the date of publication. Neither the publisher nor the authors or the editors give a warranty, expressed or implied, with respect to the material contained herein or for any errors or omissions that may have been made. The publisher remains neutral with regard to jurisdictional claims in published maps and institutional affiliations.

This Springer imprint is published by the registered company Springer Nature Switzerland AG
The registered company address is: Gewerbestrasse 11, 6330 Cham, Switzerland

If disposing of this product, please recycle the paper.

Contents

1 **Introduction** 1
2 **Tool** 5
   2.1 Testimony 5
   2.2 The Absence of Disturbance 7
   2.3 Tool 9
   2.4 The Humanities 10
   2.5 The Double Nature of the Tool 12
   2.6 Naturalism 14
   2.7 Antinaturalism 16
   2.8 Process as a Tool 17
   2.9 The Passing Time 19
   References 21
3 **The Human Science** 23
   3.1 Progress 23
   3.2 The Cycle Revisited 25
   3.3 Frazer and Forms of Thought 26
   3.4 Purpose and Use 28
   3.5 Is It Really Progress? 29
   3.6 Demands of the day 31
   3.7 The Contemporary 32
   3.8 The Actual 34
   3.9 Time 35
   References 37
4 **Change** 39
   4.1 Time as Such? 39
   4.2 The Change 41
   4.3 Manipulation, Symptom, Action 43
   4.4 Grasping Time 44

4.5	Perspective and Change	46
4.6	Absence and Goal (Effect) of the Change	47
4.7	Absence of the Goal (Effect) and the All-Embodying Nature of Change	49
4.8	Order	51
4.9	The Act and the Result	52
	References	54

5 Crack 55
- 5.1 Mandelstam 55
- 5.2 Nabokov 57
- 5.3 Cohen 58
- 5.4 Cultural Shock 60
- 5.5 Stultitia 61
- 5.6 Falsification 63
- 5.7 In a Community 64
- 5.8 On One's Own 66
- 5.9 A Couple 68
- References 69

6 Culture 71
- 6.1 Different Worlds 71
- 6.2 Not a Monad But a Human Being 73
- 6.3 Calibration of Cultures 74
- 6.4 What if There is No Calibration? 76
- 6.5 Foucault's Problem 77
- 6.6 Kmita's Problem 79
- 6.7 The Border as a Landmark 81
- 6.8 Changeability of the Border 82
- 6.9 Performative Nature of the Border 84
- References 85

7 Proposal 87
- 7.1 The Border and Culture 87
- 7.2 The Metaphor and Verification 89
- 7.3 Tension 90
- 7.4 The Rhythmical Tool 92
- 7.5 The Change and the Crack One More Time 93
- 7.6 Two Elements and Direction 95
- 7.7 With No Retreat 97
- 7.8 Reconstruction of Events 98
- 7.9 Freedom 100
- References 101

8 Conclusion 103

Index 107

Chapter 1
Introduction

Abstract This is the introductory chapter. The author presents the basic problems—representation, movement, tool, change, culture and crack—while addressing the very condition of the human being living in time (space–time). The condition is that this being is continuously exposed to various cracks where culture is offered as a remedy.

Keywords Representation · Movement · Tool · Change · Culture · Crack

This book is devoted to the topic of the representation of movement. I am referring here to various results of human work, which reconstruct, but also manipulate this particular attribute of bodies, including the human body. Since this work, which requires reflection on our own condition of entanglement in movement, is exceptionally complex, I shall begin with sketching the possibly individual context and move on to a sufficiently simple presentation of this complexity. May this context be a reference to my own life, which I attempt to represent with the help of a convenient, as I believe, figure of the river. May I not stand in front of it; may it be my vehicle. May I, therefore, move with it. Initially, I may be flowing down a swift stream, sliding over the rocky bed between a dense growth of plants filling a mountain valley. The valley is changing into hills, and they, subsequently, into lowland meadows and forests. As a result, the current is slowing down, and the stream is becoming broader and imperceptibly changing into a lazily flowing river. Finally, in front of me, I can spot the sea towards which the river is heading—it stretches far into the horizon. This is the end of my life—the course of which is unavoidable. At the same time, I become aware that the figure I have used reflects not so much life in movement, as life within the time which is reflected in the movement—and not only that. Otherwise, where does the metaphor 'time flows' come from? And it sometimes flows slowly, and at other times fast. Here, I realise how difficult it is to control the current, which is carrying me along. There are moments when I keep an eye on the order of the day or year, checking which places I am passing by and plotting this current position on one map or another. However, equally often I forget about it, turning to other matters—not so much petty perhaps as calling for my attention at a given moment, which enables me

© The Author(s), under exclusive license to Springer Nature Switzerland AG 2025
A. Zaporowski, *Crack and Culture: On Representations of Movement in Anthropology and Philosophy*, SpringerBriefs in Philosophy,
https://doi.org/10.1007/978-3-031-83422-6_1

to keep my balance during the journey. As a result, I increasingly notice that when I want to recreate the route I have already covered, it is like watching a film with many frames missing.

Such a representation may turn out to be as much a personal as a commonplace product of myself. This is because I am aware that human life, referred to time, is the object of more profound analyses encountered, inter alia, in the humanities. It is also frequently that when confronted with such analyses, the personal and the common is disregarded as lacking sufficient seriousness or depth. And if such representations are not treated with distance, they are at least categorised in scientific discourses as special cases of more general explanatory or interpretative schemes. At the same time, it is not only such discourses that are possible. Personal stories must also face a much older and often equally impactful form—that of myth and, more broadly speaking, the worldview—which regulates the social domain. Finally, there is a tendency to approach the above representation without succumbing to the pressure of the intellectual and emotional impact of scientific and mythical narration. This approach would lie in somebody's self immersing in the contemplation of an image, which would certainly require clarification of the issue of the art of representation and the recipient's sensitivity. In any case, it is possible that my story would be subject to the pressure of any of the above techniques and I assume that I would be forced to succumb to the way of storytelling with a power of persuasion much greater than I could possibly possess.

When writing about the power of persuasion, I am referring to the forms—of art, science or religion—that are old and dominating enough to make my narration useful only privately. Below, out of necessity selectively, but also in a slightly varied way, I am presenting the particular manners of talking about the human being living in principle in time, and constituting an element of a more complex system—space–time. I consider these manners tools with the help of which their constructors manipulate rather than replicate time. Without consideration of the motives behind this action, I notice that each of the manners diagnosed either obliterates or emphasizes the role of time; none of them can be limited to a single domain: be it artistic, religious or scientific. At the same time, each of them includes a persuasive load, which underlies the manipulatory nature of the tool. Of course, persuasion can have its source in the trauma experienced by the constructor, which is most often testified to in a literary work. It may also arise from the calculation of the expected results of the application of the tool. Anyway, I consider the practice of the self-identification of the human being—a social animal—with the contents produced by this tool and the formation of the relevant groups modelled by their work as significant. In this context, I suggest a reflection not only on the very functioning of such a mechanism, but also on a way of distancing from it. Simultaneously, I am conscious that an escape from any mechanism of presenting oneself and others in time is an illusion. On the other hand, I assume that a critical attitude to, inter alia, our own (including personal and private) stories is an indispensable element of maintaining a flexible relation between an individual and a social group.

I postulate that individuals develop an attitude to themselves and the already mentioned tools—their own or those of others—owing to culture. I understand culture as a system of attitudes, including propositional ones, which determine people's non-reflex actions. I refer to such actions as cultural ones, since they involve adopting a particular attitude entailing all other attitudes comprising the entire system. In this sense, a particular action is ordered; it allows the acting person and their interpreter to realise what attitude this particular person has to the series of events they are dealing with in the transient world, as testified to, inter alia, by the series of the above-mentioned tools. I call the latter cultural works, since they are the result of actions determined by a system of attitudes. I hence adopt a model in which culture is referred simultaneously to both actions and works, while the last two of them are mutually referred to one another. Culturally-acting individuals produce tools. They subsequently use—i.e. recognise—them. At the same time, there may be many such uses; they can be mutually incommensurable. The use of the tool by a constructor and the user makes a special case here—each of them may recognise a different object. On the other hand, the tool is sometimes not recognised as a cultural work—and is then useless. The reciprocally related set of action and cultural work shows the individual as a constructor and user of various mechanisms of coping with time, with the two roles being marked by potential creativity and independence. Particularly interesting is the position of the heretofore user, who can become a constructor.

When writing about mechanisms of coping with time, I am referring to the effect of an event assuming a change experienced by the individual, including any member of society. I assume that they become intuitively aware of the passage of time not when they flow through life down the swift current of their self. The change is tantamount to a discontinuation of the uniformity of experience, with one of its variants being the regularity, including repetitiveness, of such an event. In turn, such regularity may be conscious. In this context, I postulate that the mentioned tools are constructed in order to not only recreate but also process the regularity. Above, I have referred to representations, but we should not forget about the devices, which imitate the change I consider a measure of time. I am referring here to a drum reflecting the rhythm of the beating heart. It is baffling that the work of this organ consists in the reversibility of change, while the experienced testimony—the pulse—is irreversible. However, it is not only regular. The cycle of the rhythm of the drum imitating the pulse can be modulated and can affect the user in various intended ways. The same feature marks the representations which are reflected on below, which suggests that change can also be created. However, there is a special case of an event assuming change—a crack. Its nature is irreversible and irregular. As reflected in particular in artistic works, it is often a traumatic experience. Nevertheless, I assume that it constitutes a testimony, which allows the individual, especially the one desiring to be critical of their own condition, to cope with the problem of the passage of time.

I have put together crack and culture. An event assumes change understood as a relation between two incommensurable states, with the relation basically being reproducible. A crack is a type of event as a part of which the heretofore existing change finds its ending. For this reason, its consequences can be dramatic. It thus also indicates the reason behind the appearance of a series of tools—some of which mask, while others highlight it. The former creates the risk of neglecting the issue of time, while the latter—of a prejudgement of its necessary nature. Now, I refer the crack to culture to propose a model of reflection on time in the context of the road that can be travelled making the crack an indicator of the new course taken by the individual or a set of individuals, each time ordering the bundle of their actions in a new, but basically regular way. By assumption, this road is to underline the presence of time, which in this particular way is extracted from the testimony of events. I thus want to check whether the above representation of the river, with its foredoomed beginning and end, can be proved right. Therefore, I intend to put together this representation with the postulate that it is culture which allows me to move forward on the basis of the testimony in the form of the crack.

* * *

This book is my personal achievement. However, it would not be complete if it had not been translated into English by Anna Moroz-Darska. I am therefore deeply thankful to her.

Chapter 2
Tool

Abstract This chapter concerns a variety of representations of movement as tools that allow one to orient oneself in the dimension of time. It begins with an appeal to the testimony of a non-continuity of the world timely understood as exemplified by chosen academic positions. Other tools for dealing with the world in motion are then presented, where motion has a rhythm. This is the way the tools, especially the cyclically oriented ones, have been so useful in human history. These tools have a double nature: they are representational and constructional. In this sense they are used for a manipulation of what is represented. A form of manipulation is exemplified through an appeal to two opposite academic schools: naturalism and antinaturalism. They both acknowledge the notion of movement. What is missing, however, is the ordered and foreseeable nature of change, so that one's knowledge of motion may be contingent.

Keywords Representation · Movement · Time · Tool · Naturalism · Antinaturalism

2.1 Testimony

Several decades ago, Michel Foucault's essay *What is Enlightenment?* Was published (Foucault 1984) in which the author referred to Immanuel Kant's text published under the same title two hundred years earlier. The first philosopher diagnosed the second one's position as a reference to the present assuming both its distance and the task. Deciding to develop his diagnosis, Foucault described this reference as an adoption of a specific attitude he called modernity. It was to be marked by the awareness of the non-continuity (of time), irony, readiness to transform oneself, and, finally, the creation of art. The French philosopher called it a limit attitude, since he recommended that the individuals practising this attitude be at the frontier of knowledge; that they be ready to transgress it. This recommendation had a positive character. Foucault thus stood up against Kant, according to whom there exist impassable frontiers of knowledge. What the French philosopher referring to the ethos of modernity meant was not only resistance against Kant's negation of the liberal use of reason and the clear differentiation between the private and the public—he also opposed the universal and necessary against the particular and contingent. This is because he assumed that the modern attitude allowing the adoption of increasingly new attitudes to the present is attributed to the human being—a historical being, who is always an unfinished project.

Foucault's contemporary, Jerzy Kmita, referring to Ernest Gellner and Max Weber, presented the process of the transformations of developmental types of societies (Kmita 1996). An interesting feature of the Polish philosopher's concept was comparing this process with the identification of the gradual separation of what was referred to as spheres of culture. According to Kmita, culture controlled social practice, also in the cognitive sense, in two ways. Firstly, the actions of members of a given people were determined by propositions—some of which specified the goal behind the actions, while others—the way to reach the goal. Secondly, the actions were determined by the 'global state' external to the members of the people and modelled on the system of institutions postulated by classical functional anthropologists. Remembering that both these determinants were two sides of the same coin, it is worth turning attention to the first one. Kmita postulated that during social transformations—when pastoral and agricultural people were taking root, and, along with them, so too religion, which followed magic, and preceded science—two spheres: the symbolic and the technological, gradually became distinct. The former had an integrating character, while the latter was instrumental. The Polish cultural researcher not only assumed that both these spheres would become increasingly independent of each other—he also came to a conclusion that the attitude of members of a given people to propositions determining ways of reaching the goals of their actions in both spheres of culture is different. He thus showed that acting symbolically and technologically, it is possible to live in incommensurable worlds.

Foucault and Kmita differ in several respects. Firstly, the former focuses on the individual, while the latter is concentrated on the group. What is more, Foucault points out that the attitude of modernity can be attributed to anyone brave enough to distance themselves from themselves, regardless of the series of events making up the Enlightenment, which can be located in time and space and which is usually associated with the modernity as an epoch rather than an attitude. Meanwhile, Kmita reflects on transformation as a historical process, suggesting that modernity has basically the nature of an epoch. Secondly, when writing about, respectively, attitude and culture, both philosophers focus on two distinct, although mutually complementing variants of knowledge, which, according to the nomenclature of another thinker, Kazimierz Ajdukiewicz, are called a cognitive act and a cognitive result (Ajdukiewicz 1973).[1] The cognitive act in the form of, say, a reminder or an utterance, encompasses the adoption of the attitude of modernity, which makes it impossible to omit attitude as an object of analysis: the attitude is postulated by Foucault. In turn, Ajdukiewicz understands the cognitive result as propositions—in particular the ones referred to

---

[1] It should be added that Ajdukiewicz's distinction follows his teacher's—Kazimierz Twardowski—distinction between mind-dependant actions and products (Twardowski 1999).

as scientific (and affirmative ones), which however does not rule out references to propositions establishing the goals behind actions as well as actions as means: propositions are postulated by Kmita. Thirdly, as much as Foucault analyses the border—seen from one side to boot—as a challenge the individual can face, Kmita assumes a distinction (of spheres of culture) as a result of social changes. However, on the other hand—and I want to clearly underline this—both of them turn attention to what I shall refer to as a breach, disturbance, crack—testimony to the disappearance of the obviousness marked not so much by the homogeneity or continuity of the surrounding world, as by the absence of expectations or fears of its possible disturbance.

2.2 The Absence of Disturbance

However, it should be remembered that the testimony in question can be absent. There are at least three possibilities: the first one belongs to the individual and aesthetic domain, which subjects people to external sensory stimuli. This assumes the inclusion of the human individual into the infinite world as its integral part, which gives the sense of stability. When Joseph Conrad was testifying to his own experience as a sailor sailing up the unknown River Congo, which was meandering its way through the dark jungle, he wrote—through his character Marlow—about 'travelling back to the earliest beginnings of the world' (Conrad 1999: 87). The journey was turning out to be an embrace of boundless nature. This sensation made the author, who was observing it, quieten down his thoughts, succumbing to the unknown and yet soothing sensations. On the subsequent pages of his novel, Conrad pointed out that a travelling individual—whether a native hired for work for several months, or a European—can have no idea of time (Conrad 1999: 96). He thus said that the experience of the boundlessness of the jungle allows one to lose orientation, to allow the omnipresent colours, shapes, sounds, and smells to make up one's self. Nevertheless, the experience in question was temporary—after all, after a moment of contemplation, he had to go back to his tasks. Anyway, in periods of its intensification, the world seemed to be a uniform system of overlapping sensations one could not step beyond.

The second possibility is already seen in the social, but still aesthetic domain, where stability assumes repetition. It results from Conrad's observations—while being a writer, he simultaneously turned out to be an ethnographer experiencing the alien in the most basic sense of the word. This is because his novel is a record of the world which had not yet been colonised—sub-Saharan Africa at the end of the nineteenth century. The eponymous heart of darkness, with the darkness meaning both the all-encompassing forest, and the all-encompassing soul, is perceived when the author hears the sound of a drum. It turns out that the rhythm reproduces the rhythm of the heart. Here, being a part of nature does not lie in immersing oneself in it, but in discovering it in oneself. This is because nature is not so much boundless, as the forest, but pulsating, as one's own body. What is more, this sensation is both experienced in a natural way, and as a result of human manipulation. Normally associated with nomads—and such peoples used to live in the part of Africa in question at the time—the practice of imitating the rhythm of the body by an instrument can be understood as a way of evoking the sense of continuity, which usually accompanies individuals, harmonizing their pulse with the course of events. The pulse is cyclical—circular. Hence, it is closed, which evokes the sense of balance. The drum, which imitates it, allows not only a single individual, but the entire community to refer to the mobile world as a permanent whole. If the previous possibility assumed the projection of the homogeneous world onto the self, in this case we are dealing with the projection of the pulsating self onto the world.

Finally, the third possibility is still a part of the social and aesthetic domain, where stability is possible owing to the representation, which immortalizes not just the repetition, but also other forms of movement. So far, the reader of Conrad's novel has had in front of them a nomad, who basically reproduced the order of the world, including that of their own body, on an ongoing basis. Examples of the above may include the listening to the sound of the drum and reacting to it. The typology of societies developed by the above-mentioned Gellner also covers shepherds and farmers, who, when communicating, not so much imitated as represented an object, with the practice assuming an idea of the inevitability of the succession of one object after another, with the succession often being linear. In turn, the idea might have appeared as a result of the use of a tool which perpetuated movement. What appears in this context is a special form of coping with the world, including oneself—myth. Perpetuation may suggest the use of the script, since, inter alia, when indicating myth as a means used to have power over the non-continuity of the world, Leszek Kołakowski gives such examples of books as *The Republic* by Plato, *The Torah* and *The Bhagavad Gita* (Kolakowski 1989). However, mythical parables can be transmitted orally—and the systemic nature of such practice can be exemplified by the example of the ancient Celts. Hence, I understand perpetuation as a generalisation of direct experience in the form providing the sense of the continuity of the world. What is simultaneously interesting is not only the said inevitability of the sequence of one thing after another, i.e. the ensuring of stability owing to the necessary (implicitly: closed) nature of the chain of events, but also the often cyclical nature of the reproduction of this representation in the customary form.[2]

---

[2] Of course, customs do not have to have a cyclical form to have a regular character. May the above be illustrated by the obligation to shake hands when meeting somebody we know. Nevertheless, this cycle very distinctly underlines the reproductive nature of the custom. When writing about the custom further on, I shall have in mind its cyclical form.

2.3 Tool

The above typology of possibilities of the absence of any expectation of the disturbance of the continuity (homogeneity) of the world, has so far been presented from the point of view of the individual experiencing it. However, it calls for complementation when talking about the human individual as an animal reacting to stimuli in an ordered manner. The first possibility (that of every individual being affected by external sensory stimuli) is out of the question, since a reference to the aesthetic domain does not necessarily presume a reaction in the form of the manufacture of a tool.[3] When Geertz put together various perspectives—in simplified terms, sets of (propositional) attitudes allowing a methodical answer to the experienced events[4]— he distinguished the aesthetic perspective, which was marked by—according to Susanne Langer whom he quoted—'(…) disengagement from belief—the contemplation of sensory qualities' (Geertz 1973: 111), generated by works of art (poems or sculptures). It is in this way that Marlow's immersion in the primeval forest can be interpreted. Conrad's novel's hero's observation of the world without any practical, critical or dogmatic purpose did not result in the production—even for himself—of any tool for the ordering of this world. This is why such a creature may have no sense of time. The situation with the two remaining possibilities is different. Obviously, we are still dealing with an individual experiencing the continuity of the world, but this continuity inevitably clashes with the experience of the movement. This experience is modelled or filtered by a relevant tool.

Without any doubt, people using such a tool—a drum or a mythical parable (also the one reproduced cyclically)—may represent the Geertzian aesthetic perspective. It is not necessarily so in the case of the producer of this tool. When analysing the development of human communities, Edward Hall distinguished a number of communication revolutions, three of which are particularly interesting in the context of the present reflections. The first revolution, which took place about 100,000 years ago, consisted in the appearance of language as an ability to speak. After ca. 90,000 years, the world experienced the second revolution, as a part of which 'some rather bright but not very well adjusted types' (Hall 1998: 56) discovered that language can be an object of reflection (as a part of metalanguage). Finally, after the subsequent several thousand years, the third revolution—the recording of speech—took place. It would seem that the creators of the drum—the nomads—experienced the first revolution, while the creators of myths—shepherds and farmers—went through the second (if they cyclically transmitted parables orally) or the third one (if they read them out cyclically). Does this, however, mean that the first ones were not necessarily creatures reflecting—vide the appearance of the metalanguage—on their own practice (also

---

[3] I am not reflecting here on the possibility of the creation of a tool for contemplation—as a part of the aesthetic perspective—which is tantamount to the avoidance of the ordering of the reality I am reflecting on. Fragment's of Conrad's novels are one such tool.

[4] Geertz was indeed mistrustful towards such constructs as that of attitude, and that is why he used the term 'perspective'. Here, however, I am by necessity simplifying, while discussing Geertz's standpoint elsewhere (Zaporowski 2021).

the linguistic one)? Perhaps. After all, Hall writes about the first users of language as the creatures acquiring knowledge 'without awareness' (Hall 1998: 56). However, if linguistic competence is considered to be an ability to quantify, testify to the predicate of satisfaction, and refer to propositional attitudes—in a nutshell, to abstract—then the absence of a testimony to action at the metalinguistic level should not be an obstacle to acquiring an ability to react to experience in an orderly way, but at the same time creatively.

In this sense, the drum and the parable on Adam and Eve are equally good candidates for a tool the user of which does not overstep the aesthetic perspective. As far as, then, people can be considered creatures able to contemplate the world, the effect of the use of the tool, which copies or represents this world (or its element) can also be the source of such contemplation. However, this use—to refer to the pragmatic domain, which is indispensable in this case—testifies to the fact that in the context of the quoted example, at least the creators of such a tool have the sense of, inter alia, time. This is because they can recreate it. What do I mean by this? Although a lack of movement is reflected on, it is very difficult to grasp time. However, once movement appears, time, being its reflection,[5] becomes easier to grasp. Albert Einstein, whom Hall referred to, stated that 'time is what a clock says and that anything can be a clock: the rotation of the earth, the moon, and other rhythms' (Hall 1998: 66). The clock such as the drum (or the heart) assumes a relatively clear identification of the rhythm and the possibility to manipulate it by its creator (or discoverer). In this sense, this identification is a testimony to the above-mentioned ability to abstract— the skill to establish the relations between the whole and its parts.[6] The example of the myth is more complex, since on the one hand it perpetuates (generalizes) the not necessarily linear, but necessarily closed form of the sequence of events, while on the other hand—it is a component of a special clock—the already mentioned custom— measuring time just like the drum. This is why it is so significant to differentiate between the tool and its use.

2.4 The Humanities

As already stated, a mythical parable differs from the drum by its having a representational (rather than an imitative) and generalising nature. However, this does not mean that it does not refer to time, as testified to by the already mentioned inevitability of succession. Also, it turns out that it may have more properties than just this one. May the above be testified to by a set of analyses of this still archaic tool, carried out by researchers representing the humanities—a field of knowledge stretching from cultural and social anthropology and linguistics to philosophy, which

---

[5] Interestingly, despite movement being observable, and time—not observable, the same predicates can be used in both cases—for example 'uniform'.

[6] Contemplation does not allow such an identification; it only enables the establishment of the relation between parts, but not between any of them and the whole.

2.4 The Humanities

in the context of the transformation represented by the Enlightenment understood the individual not only as a body, but also as a social and cultural creature. From among these researchers, in the context of the tools under discussion, it would be worthwhile to mention the fairy tale researcher,[7] Vladimir Propp (1968), and the myth researcher, Claude Lévi-Strauss (1955). The former emphasized the linear (one-dimensional) character of parables in 1928, while the latter diagnosed their geometrical (two-dimensional) character in 1955. According to Propp, fairy tales are divided into parts—functions—describing the actions of their protagonists, which are significant for the course of the plot. There are 31 such functions and although not all of them are present in every fairy tale, each sequence—events succeeding each other—composed of a given number of functions, testifies to the same scheme, which assumes inter alia preparation, the generation of a problem and its solution.[8] In turn, Lévi-Strauss pointed out that at the base of the linearly presented narration composed of mythemes—equivalents of Propp's functions—there is a structure—a pair of binary opposites between the mythemes.[9] It testifies to two intersecting lines—a horizontal and a vertical one—which made myth a tool of a ceaselessly updating balance. Hence, he noticed that a myth is not only a representation of the necessary sequence of events—the discovered structure indicated that it has its own rhythm, that it is based on repetition.

I consider it intriguing that when analysing various forms of social practice—including the above-mentioned tools—anthropologists and other representatives of the humanities created new forms themselves and, using them, influenced the sense of time of their new listeners (readers). Foucault and Kmita are such constructors, while their theses, which have been mentioned above, refer to two alternative scenarios of the mechanism of the organisation of movement. The first one assumes a system consisting of two, while the second one—of three—elements. Foucault's model is obviously about transmission from one state to another, which is to be made possible by the attitude of modernity. It is assumed that not everyone is able to adopt such an attitude. Hence, Foucault admits not only the possibility of the individual working through themselves by stepping beyond a specific border—but the French philosopher constructs a model of two types of people: modern and anti-modern ones. The former are mobile and creative, while the latter—static and conservative. In view of the above, the listener (reader) of Foucault's narration must put together the system of transmission from one state to another and the system of the two opposite types of

---

[7] The relation between the fairy tale and myth is complex. For instance, Propp considers the fairy tale a derivative of myth, while Lévi-Strauss believed that these forms are distinct, although complementary. The fact that myths and fairy tales are closely interrelated is testified to for example by Lévi-Strauss's analysis of Propp's concept (Lévi-Strauss 1960).

[8] This scheme is based on the logical consequence of A $\rightarrow$ B.

[9] I wish to stress that Propp assumed that some functions come together making pairs, although a pair of functions is not tantamount to a pair of binary opposites. In this sense, it can be concluded that to some extent, he anticipated Lévi-Strauss's position. On the other hand, in the case of both analyses, movement is ordered in a different way. Here, Propp uses a convention underlining time, while Lévi-Strauss—a structuralist one, which blurs time. This is going to become clearer further on.

people. In the last case, we are dealing with a reference to the relation of opposition, which is stabilizing. The specificity of Foucault's standpoint simultaneously lies in the procedural understanding of this relation. In consequence, the reader (listener) obtains a two-dimensional model, as a part of which the diversity determined by the border is accompanied by the diversity determined by the attitude.

Kmita's model is also two-dimensional, showing in what way the domain of social transformations intersects the domain of the spheres of culture and propositions respected as a part of them. Here, however, the two-element system intersects the tripartite one, where the transformation of a single social type into another is integrated with the parting of the spheres of culture according to a relevant key. The first transformation—of the nomadic into the settled—is accompanied by a preliminary parting consisting in a relative domination of the symbolic over the technological. The second transformation—of the settled into the industrial—is equivalent to the final parting, resulting in a relative independence of both spheres. On the one hand, both these models testify to the process. However, on the other hand, every process is composed of specially connected elements. As far as the creator of the tool should be aware of both these aspects, its user may fail to perceive the last one, focusing on the homogeneous representation of the sequence of transformations. Anyway, my remark indicates that we should look at a given tool just like at a medal—from both its sides. The first side—representational[10]—is usually available to the user, possibly homogeneous and fulfils its purpose, when the very tool seems to vanish from the field of vision, making it possible to see the world it has ordered. The second side—constructional—controlled by the creator, is possibly differentiated, allowing an increasingly better adjustment of the tool to its role owing to analytical measures. This double character of the tool results from the understanding—at least by some of the tool constructors—of the relativity between the comprehensive and the divisible.

2.5 The Double Nature of the Tool

It is possible to provide at least three examples of the double character of a given tool. The first one refers to Willard Van Orman Quine, who analysed the logical and epistemological status of language. In particular, the American philosopher asked in what sense sentences are theory-laden (Quine 1990). What he meant was the observation sentence, which is usually analysed as an incidental creation referring to the contingent state of things in the world. According to Quine, such a sentence has a double nature. From the point of view of its reference, i.e. the mentioned fact, it constitutes a whole. When analysing a person learning a language, including an ethnographer finding themselves in an alien world, he compared them to a child and called such a sentence a holophrastic one. Examples include 'Mummy', 'Milk', or 'Wolf'. They are single words used by children. From the formal point of view,

---

[10] Foucault and Kmita propose a tale which is a representation. In the case of the tool in the form of the drum, we would be talking about imitation.

2.5 The Double Nature of the Tool

they are sentences rather than names or predicates, as they can be developed to make a complex form such as 'This is Mummy', 'This is milk', etc. According to Quine, an exemplary ethnographer taken as an example learns to use not only such formulas, but also their clusters—say, when referring to mum and milk at the same time. However, they are still at the occasional level—they are not yet a competent user of the language. The ethnographer acquires competence only when they can use lawlike sentences (also referred to as categorical observation sentences), testifying to their ability to use the predicate of satisfaction, the generalised form of which is 'Whenever A, then B'.

The subsequent stage involves the liberation from observation, and thus—occasionality. Taking into account the generally adopted (in the cognitive context) division of sentences into observation and theoretical ones, it might seem that only the latter (to which the lawlike sentence is an approximation—hence its name) guarantees the achievement of a level of continuity, i.e. liberation from occasionality. In this place, Quine states that the double nature of the observation sentence lies in the fact that apart from its reference to the state of things, its parts can be referred to each other. For instance, 'Milk' may be a compound of 'White and warm'. However, the already mentioned turning of attention to parts requires a simultaneous perception of the logical conjunctions which connect them. These compounds are not empirical and hence indicate the burdening of the observation sentence with theory. The second example is similar to the first one. When analysing the issue of actions as instances of events, Quine's student, Donald Davidson, referred to an utterance as a physical event. As a whole referred to other events, it has a processual character. However, when it is analysed from the point of view of its structure, it will turn out that there is 'a relation between a sentence, a person, and a time' (Davidson 1967: 319). In this context, losing its performative nature, it will obtain the status of a cognitive result, which I have mentioned above when referring to Ajdukiewicz.

The third example can be found in Geertz's analysis of religion as a cultural system (Geertz 1973). This American social anthropologist distinguished, inter alia, the symbol, and called its meaning a conception. In this sense, a given physical object or a physical event could be understood by community members as a product of their culture. According to Geertz, the symbol had a double nature: it was a model for something, and a model of something. The former was used for manipulating the world, just like people manipulate it, when building a house. The latter was used for manipulating semantic relations, just like people manipulate them when creating a plan of the house, calculating the cost and establishing the sequence of works. We may use the following sentences as examples: 'mythological figure materializing in the wilderness' and 'the skull of the deceased household head hanging censoriously in the rafters' (Geertz 1973: 95). On the one hand, the symbol as a model for something affects the moods and motivations of people—members of the community—while on the other, it expresses abstract relations, which allow to identify these people in the order of the cosmos. Just like the two previous sentences, also this one shows that we may look at a given object or event—be it a skull, a painting or an utterance—from two sides. As a tool, this object or event will be a comprehensive representation (or imitation) exerting its impact on the user, while simultaneously being a complex system, with the creator having control over its order. Although not just the creator.

2.6 Naturalism

The bringing of a new element—the manipulation of differentiations—into play is historically justified. This is because I assume that Foucault's and Kmita's examples express attempts at the positioning of reflection on the human individual in the structure of the human sciences, which have themselves been undergoing transformations. It was in their framework that, in the nineteenth and twentieth centuries, a dispute between naturalism and anti-naturalism (also methodological one) took place.[11] One of the issues of this dispute was the method of ordering the social and cultural reality. Representatives of the first approach referred to the heritage of the natural sciences, where explanation was the desired method. For example, the German-American philosopher of science, Ernest Nagel, concluded that explanation could be defined as an answer to the question 'why?' As a result, he presented four types of scientific explanation: deductive, probabilistic, functional (teleological) and genetic (Nagel 1961). From among them, the first three assume ordering through generalisation; the genetic explanation concerns in principle the consequences of single facts. Regardless of whether it is immobility or movement which is the object of explanation, it consists in the representation of the relation of dependence of the instance which is being explained—the explanandum—on the instance which explains—the explanans. A specification of any of these states turns out to be secondary. This is because of the very relation and the way in which such a representation is performed.

The relation under discussion—or, speaking in more detailed terms, the rule on which it is based—is abstract, and this is so regardless of whether the explanans is composed of general laws or hypotheses, as in the case of testifying to causal relations, or whether the explanans identifies a feasible goal, which is taken into account in the case of functional relations. This relation takes place outside time and space. In turn, representation is specific, and hence requires a reference to a dimension. Interestingly, this representation privileges the reader's (listener's) ability to imagine spatial relations, although they hamper the reflection of movement. Let me refer here to the example of empirical testimony to the reflective reasoning—the aforementioned rule—which takes on a graphic form. *Modus ponendo ponens* is a rule of deductive reasoning lying at the basis of, inter alia, the explanation of causal relations,

---

[11] On the one hand, the term"naturalism" can be found in Carl R. Popper who discussed the problem of scientific method (Popper 2002). He identified naturalistic approach with positivism (and induction) while opposing it with the conventional approach (and deduction) he opted for. The approach under Popper's attack well illustrates the is/ought problem articulated by Hume (2000). According to Hume, one must not derive the normative propositions from the descriptive propositions. On the other hand, I find the roots of the opposition between naturalism and anti-naturalism in a foundation of a division between *Naturwissenschaften* and *Geisteswissenschaften* discussed on the following pages.

2.6 Naturalism

maintaining that conclusion (B) follows necessarily from the logical conjunction of the premises (A → B) and (A). It can be interpreted in time in such a way that in front of the reader's eyes, one character appears after another. However, this interpretation normally does not take into account the possibility to grasp the reversibility of the order between the relevant elements (A and B).[12] In this sense, the spatial representation seems to be more relevant than the temporal one. Simultaneously, let us not forget that naturalists most often defined explanans as the general or the typical, and in such cases time tends to be disregarded.[13] Let us consider, for instance, that in Nagel's typology, only the genetic explanation necessarily requires taking into account the temporal dimension, as it concerns the relation between the particular facts. Therefore, remembering about the relation between the extrasensory nature of explanation and its sensory representation, it should be concluded that the naturalistic approach of representatives of the humanities allowed a more frequent construction of both generalising and static images.[14]

This general and static nature of representations of the order of facts becomes in particular visible when we notice that the said causal and functional relations may be reduced to the relation of conditionality (Zaporowski 2018). For instance, if somebody asks why plants have flowers, they will hear as a part of the teleological (functional) explanation that this is about the provision of the continuity of life. In this place, we may notice temporal relations, since first a given flower must be pollinated, and then a seed must be developed to finally sprout and make it possible for a new plant to grow. However, it is only when we have made the generalisation, recognising that having flowers is a means for the plant to reach its purpose—the continuity of life—that we may notice that logical relations may also be involved. This is because this means will be considered a sort of a sufficient condition, while the purpose—the necessary condition, with the absence of balance between the two elements testifying to the imperfection of the order being sought for. This is what Lévi-Strauss had in mind when he wrote that '(...) causality (and the succession in time which it implies) must be rejected (...) in favour of a functional correlation between the two phenomena' (Lévi-Strauss 1952: 47). The French ethnologist did not refer to the functional relation, an example of which I have presented above.[15]

---

[12] What I have in mind is a situation when the said deductive rule is replaced with a reductive rule: (A → B) ∧ B → A. The order between A and B is reversed here—in the sense that as a part of the reductive rule, B is a smaller premise, while A is the smaller premise in the deductive rule. Let me add that these rules are not commensurate, but manipulation with spatial relations reflects the relation between them better than the manipulation with temporal relations. This is so because they are simultaneous alternatives rather than possibilities following one another in sequence.

[13] However, the general or typical remains regular, since this last property can be understood independently of time. As an example, we may use the above-mentioned categorical observation sentence by Quine.

[14] In this text, I am also presenting—having in mind various genetically- and evolutionally-oriented researchers—naturalistic concepts taking into account the dimension of time in the explanans.

[15] The causal relation criticised by Lévi-Strauss assumes the presence of the temporal dimension, although it should be added, against the French ethnologist, that the cause may be not only earlier, but also concurrent with the result. The essence of the issue lies in the cause constituting in principle the sufficient condition of the result. At the same time, the presence of the dimension of time also

What he wanted to discuss was the variant in which each of the phenomena testifies to the sufficient and necessary condition in relation to another. Hence, the finding of a (logical rather than actual) balance between the examined elements is the basic goal of the researcher. In turn, the appropriate vocabulary testifying to the potential balance is the vocabulary as a part of which relations express the possibly general and simple conditions.

2.7 Antinaturalism

It is not incidental that such a variant of the functional relation—the variant available owing to the use of the said vocabulary—was analysed by Lévi-Strauss's predecessor, Émile Durkheim, the creator of the French school of sociology (Durkheim 1984). He understood it as a relation of correspondence, which testifies to both the sufficient and the necessary condition. Among other things, the reader learnt that the type of society with mechanical solidarity is a function of the principle of similarity, while the type of society with organic solidarity is a function of the principle of the division of labour. The specific level became available only after acknowledging that both relations were testified to by the law in force—penal and restitution (civil) law respectively. However, at the turn of the twentieth century, the naturalistic variant of the humanities had its alternative in the form of the antinaturalistic approach practiced inter alia in Germany. It was at this time that the division into *Naturwissenschaften* and *Geisteswissenschaften*, based inter alia on the methodological criterion, was popularised. In this context, nature and spirit were convenient figures, which were to testify to whether, following in the steps of Immanuel Kant, a given object of analysis was subject to, accordingly, the general laws or not.[16] The 'spirit sciences', also referred to as different historiographies or the cultural studies, emerged when their followers recognised claims of the natural sciences as ones reducing the individual—also an element of nature—to an adaptive system based on the action of the reflex arc.

I believe that one of the key differences—in the methodological sense—was the one identifying the former sciences with explanation, and the latter ones with understanding. A very special typology of understanding was developed by Wilhelm Dilthey. According to the German philosopher, the object of understanding was the manifestations of the spiritual life of the individual, being at the same time a community member. The range of manifestations stretched from gestures to works of art. Hence, this object was principally empirical (particular). The understanding

---

assumes a variant of the functional relation I have presented above. The variant of functional relation proposed by the French ethnologist disregards time, referring to the logical relations establishing the relevant condition.

[16] Here one can introduce the nomothetic/idiographic distinction. Since the 'spirit sciences' may have not acknowledged the general laws, they may have been accused of approaching the relevant object of study in terms of its particular yet contingent properties. My hunch is that one can find its equivalent in the type/token distinction.

was to consist in the identification of meanings and related attitudes expressed in the products of the acting people—social and historical animals. Dilthey, referring to his predecessor, Friedrich Schleiermacher, recommended the use of the method of the hermeneutic circle, allowing cyclical references of parts to the reality. Taking into account the fact that the written work of art was a special object of analysis, we may say that this procedure testified to the understanding person's ability to abstract rather than solely to contemplate. This procedure was not to be ever finished and had, simultaneously, two extreme forms. On the one hand, it was directed at the understanding of creatures that were possibly near to one another. Accordingly, Dilthey postulated the so-called elementary understanding consisting in the recognition of the hidden principally on the basis of an analogy to our own lived experience, i.e. in the avoidance of generalisations.

However, this sort of ability failed when people wanted to unriddle the sense of the manifestations of the life of the ones with whom they did not share that experience—the creators of past times and communities. In such cases, it was higher understanding which was at play. It was not accidental that Dilthey wrote that 'orderly and systematic understanding of fixed and relatively permanent expressions of life is what we call exegesis or interpretation' (Dilthey 1972: 232). What he had in mind was a situation when the understanding person is enforced into the 're-experiencing of a nexus of lived experience' (Dilthey 2002: 235) of a specific structure rather than an actual figure.[17] This called for distancing oneself and reflection rather than projecting one's own manifestations onto the figure under analysis. Hence, interpretation consisted in referring to inductive reasoning, i.e. abstracting properties from their carrier. Dilthey's understanding of induction differed from its classical understanding, as it involved a compilation—*vide* the relation between a part and a whole—of various elements, while the former assumes a set of homogeneous elements. Nevertheless, the 'higher' variant of the understanding testified to the generally extra-intuitive nature of the procedure of the ordering of the examined reality. On the other hand, the processual—implicitly infinite—nature of interpretation was to protect the researcher against an excessive reduction of individual manifestations of life to a specific type.

2.8 Process as a Tool

The dispute of the naturalists with the anti-naturalists—with the number and quality of their positions much exceeding the ones presented above—disclosed a significant aspect of results of the use of the research tool by the anti-naturalists. I am referring here to the reflection of the processuality of interpretation. Durkheim's example shows that a considerable group of early modern researchers privileged an idea of motionlessness. The functional relation between two elements, understood as

---

[17] What I have in mind is a situation when an analysis concerns a figure we have not had contact with. Therefore, such a figure would turn out to be a research construct.

their being a reciprocal sufficient and necessary condition, allows to understand the reality under analysis as stable and balanced, as testified to by a representation of the spatial relation where A is shown next to B. In turn, the example of Durkheim's contemporary, Dilthey, enables us to represent the world in motion. In this case, the particular elements are referred to the whole through the temporal relation, where A stands before B. However, the German anti-naturalist referred to the figure of the circle—the hermeneutic circle—rather than the line. Nevertheless, it is not a geometrical figure which is important here. This is because a reference to the latter may suggest a vicious circle, i.e. an incorrect understanding as a part of which explanandum assumes explanans, which in turn assumes the explanandum. The figure of the spiral turns out to be better, although in its case the indication of processuality is not forgone. Only the use of an image of the temporal relation in the form of a tale—a series of successive representations—turns out to be the most convenient. Still, the figure of the spiral proves very helpful.

This is because it enables the understanding of what differs the tool used by, inter alia, Dilthey, from myth as a tool. The understanding is open; it develops incessantly. Similarly to the drum, myth has a closed nature. Firstly, not every myth is linear, and I am not thinking here about its structuralist analysis. However, assuming that this is about such a case, we should indicate an identification of the end or the purpose which the protagonist of the myth reaches. It is not necessarily about a happy ending; a promise the fulfilment of which can also be ceaselessly postponed may also be at play. This end (purpose) also makes myth a closed whole. Secondly, the above type of myth can be developed into a circular myth as a part of which a promise is kept. Let us consider the example of the parable of Adam, whose sin is redeemed by Jesus.[18] Here, we can see clearly how the end is tantamount to a return to the beginning. Thirdly, the linearly- and the circularly-oriented myth can be activated cyclically, also as a part of the aforementioned custom. The latter is a clock infinitely measuring the same amount of time, with the lapse of time being an illusion. What is symptomatic here is the reading of myth by another anti-naturalist, a representative of the Marburg school of neo-Kantianism, Ernst Cassirer (1944). He examined symbolic forms which were to testify to the exceptionalness of human life—just like Diltheian expressions. Referring inter alia to myth, he put together this form of narration with the scientific narration and showed that the latter is analytical, while the former—synthetic.[19] Cassirer pointed out that the so-called prehistoric humans not only treated the world as a whole, but considered themselves a part of this whole and in this sense expressed a 'solidarity of life'. This whole had a performative nature, with the intellect and feelings mutually interrelated. It is inter alia for this reason that these people could simultaneously distinguish components of the world and feel an emotional bond with them.

The understanding of the world simultaneously as a whole and through its parts was possible owing to the people's transformations called metamorphoses. Here, the

---

[18] The coming of the Messiah is deferred by Jews, while Christians recognise him as Jesus.

[19] In Cassirer's studies, this opposition indicates the difference between the domain of the intellect and the domain of the feelings (sympathy).

2.9 The Passing Time 19

topic of the synthetic approach to the world returns. The world was One, which took on different forms. These forms transformed into one another, always testifying to the One. Let me add that the above does not only concern the prehistoric humans. Let us imagine the first Christians, who assume that they are burdened with original sin and in this sense—death. Still, they believe that they will rise from the dead owing to Christ. This transmission from the state of death to the state of life is one of the metamorphoses referred to by Christian churches—communities—using the specific clock—that of Easter. Christians do not have to control the relation between the parts and the whole, while at the same time they may undergo cyclical transformations involving myth as the catalyser. Simultaneously, not forgetting that Cassirer, following the footsteps of Dilthey, considered the lasting of man in the world developmentally, it is worth pointing out that, succeeding magic, religion contained many of magic's elements, including the possibility of referring to metamorphosis. On the other hand, remembering in particular Kmita's position concerning the parting of two spheres of culture at the religious stage, we may notice the diversity of attitudes of the followers of a given religion to its magical heritage. For instance, the attitude of Christians to the act performed by Jesus during the Last Supper is an interesting case. Some of them acknowledge that wine and bread transformed into blood and body, while others see the symbolism of blood and body in the wine and bread.

2.9 The Passing Time

As much then as myth and the drum are means allowing the recipient to liberate themselves from time, the Diltheian interpretation testifies to its passage. This happens as the researcher goes further from the starting point, putting together the whole and the parts in increasingly diverse—i.e. increasingly understandable—configurations. Therefore it would seem that in comparison with myth (and the drum), research analysis is troublesome, as it faces the recipient with potential cognitive disturbance. Time passes, which potentially does not allow a return to the beginning, while posing questions about the unexpected, the new. Cassirer also acts in this way, presenting people wrestling with the indifferent world in motion. Along with language, art and science, myth is the only way to free oneself from this indifference, while man himself is the creator of these symbolic forms, with creativity being understood processually. Can we look at this creativity in a different way? I shall answer this question, recalling Ajdukiewicz's division of forms of knowledge. The first one was a cognitive act (such as an utterance), while the second one—a cognitive result (such as a proposition), with the former processual, accentuating time, and the latter based on relations and overlooking time. Let us therefore refer this differentiation to the dispute between naturalism and methodological anti-naturalism concerning myth. It might seem that Lévi-Strauss should refer to this object in the categories of the cognitive result, while Cassirer should analyse it in the categories of the cognitive act. However, the issue turns out to be more complex than this.

Firstly, we cannot deny that the French structuralist analysed myth as an abstract structure, with its parts—mythemes—being two-dimensionally (horizontally and vertically) connected with the help of, accordingly, metonymical and metaphorical relations. In this context, the mythological thinking related to myth would lie in the intermittent passage from one relation to another. The logical thinking—oppositional to the latter—would then be involved only in the horizontal relation, as can be seen by returning to the above-presented rule of deduction. In this sense, the perspective adopted by Lévi-Strauss allowed him to refer to myth as a cognitive result—something lasting independently of the user, also in the context of its recording. In turn, the German researcher of symbolic forms referred to myth as a performatively understood way of domesticating the world understood as an action. Let us point out that this is not only, and not above all, about the text, but about the tale to be listened to. What is more, the told content is understood performatively, pragmatically—in the sense that it makes the listener take the appropriate steps such as becoming someone or something owing to the relevant transformation. In this sense, myth becomes a cognitive act or a series of such acts. The difference between Lévi-Strauss and Cassirer is also noticeable when the system of metonymy and metaphor is put together with metamorphosis. The former testifies to the relational, while the latter—to the performative.

However, secondly, the symbolic form as an object of analysis liberated from its user (for example the listener) can also be interpreted as a cognitive result—a tool designed to harness the world. I am referring to an object, which is differentiated from other objects as a part of an analysis. This is because we are free to ask what differentiates myth from religion and the latter from art. Obviously, from the point of view of the said user, all symbolic forms may maintain their processual status, but as an object of analysis, they lose it. It is one of the consequences of the consideration of the double character of the tool, as testified to by Quine's, Davidson's and Geertz's examples discussed above. In this sense, it turned out that when writing about symbolic forms 'in general', Cassirer acknowledged them as cognitive results. In turn, when Lévi-Strauss used the oppositionally interpreted relation 'stationary history/cumulative history' while analysing differences between numerous peoples, he came to a conclusion that there is no community with a static history as such (Lévi-Strauss 1952). According to him, history was relative. A given people turned out to be more or less cumulative in relation to another one. Therefore, it appears that he also perceived the processual aspect of social life.[20] It results from a comparison of these two issues that Lévi-Strauss and Cassirer adopted opposite positions and that they changed them depending on how they interpreted the individual—subjectively or objectively—as a creator or as a user of the time-controlling tool. However, we must not forget that these two aspects harmonise with each other. This is also how I understand my opening comparison of Foucault and Kmita. However, let me

---

[20] It turns out below that Lévi-Strauss criticizes social evolutionists. However, once we learn that the cumulative nature of the people increases along with the broadening of its relations with other people and decreases the more it is isolated—we are able to suppose that the French ethnologist assumes a process of transmission from the simple to the complex. He thus becomes an evolutionist.

point out that as much as in the methodological sense the cognitive result successfully reflects spatial relations, the cognitive act—including the spirally-interpreted understanding—is not an exact measure of time. Although I know as a part of the act (understanding) that I am no longer in the state I was in previously, my knowledge may be contingent. Below, I shall reflect on one attempt at coping with this problem.

References

Ajdukiewicz K (1973) [1949] Problems and theories of philosophy (trans: Skolimowski H, Quinton A). Cambridge University Press, Cambridge
Cassirer E (1944) An essay on man: an introduction to a philosophy of human culture. Yale University Press, New Haven
Conrad J (1999) [1899] Heart of darkness and other stories. Koenemann, Koeln
Davidson D (1967) Truth and meaning. Synthese 17:304–323
Dilthey W (1972) [1900] The rise of hermeneutics (trans: Jameson F). New Literary History 3(2):229–244
Dilthey W (2002) [1910] The formation of the historical world in the human sciences. Selected works, 3 (trans: Makkreel RA, Oman WH, Scanlon J). Princeton University Press, Princeton
Durkheim É (1984) [1893] The division of labour in society (trans: Halls WD). Palgrave Macmillan, London
Foucault M (1984) What is enlightenment? In: Rabinow P (ed) The Foucault reader. Pantheon Books, New York, pp 32–50
Geertz C (1973) Religion as a cultural system. In: Geertz C The interpretation of cultures. Basic Books, New York, pp 87–125
Hall ET (1998) The power of hidden differences. In: Bennett MJ (ed) Basic concepts of intercultural communication. Intercultural Press, Boston, London, pp 53–67
Hume D (2000) [1739] A treatise of human nature. Clarendon Press, Oxford
Kmita J (1996) Towards cultural relativism with a small 'r' (trans: Kwiek M). In: Zeidler-Janiszewska A (ed) Epistemology and history. Rodopi, Amsterdam, Rodopi, pp 541–614
Kolakowski L (1989) [1972] The presence of myth (trans: Czerniawski A). The University of Chicago Press, Chicago
Lévi-Strauss C (1952) Race and history. UNESCO, Paris
Lévi-Strauss C (1955) The structural study of myth. Journal of American Folklore LXXVIII(270):428–444
Lévi-Strauss C (1960) L'analyse morphologique des contes russes. Int J Slavic Poetics Linguist 3:122–149
Nagel E (1961) The structure of science: problems in the logic of scientific explanation. Routledge & Kegan Paul, London
Popper KR (2002) [1935] The logic of scientific discovery. Routledge, Abingdon, New York
Propp V (1968) [1928] Morphology of the folktale (trans: Scott L). University of Texas Press, Austin
Quine WVO (1990) Pursuit of truth. Harvard University Press, Cambridge
Twardowski K (1999) [1911] Actions and products. Some remarks on the borderline of psychology, grammar, and logic. In: Brandl J, Wolenski J (eds) Kazimierz Twardowski. On actions, products, and other topics in the philosophy. Rodopi, Amsterdam, Atlanta, pp 103–132
Zaporowski A (2018) Action, belief, and community (trans: Moroz-Darska A). Peter Lang GmbH, Berlin
Zaporowski A (2021) An anthropological perspective. A particular case. Przeglad Kulturoznawczy 2(48):319–338

Chapter 3
The Human Science

Abstract This chapter deals with an intellectual formation rather than a discipline, so the term "science" used in title is used in the singular form. In the main text, however, the classic expression "the human sciences" is used. Here we have a new form of tools to represent movement. They are based on the linear mode of the succession of events where the goal of this succession or the absence of a goal are the variants of this mode. Here we have a number of scholars who postulate an unrolling yet complex process of, first, social evolution and, second, mutation. All use the figure of a triad, which allows one to imagine not only a movement but also an irreversible direction in which this movement takes place. I then show how to finally expose time evidenced by the movement in question.

Keywords Anthropology · Evolution · Mutation · The actual · The contemporary

3.1 Progress

When Lévi-Strauss considered the above-mentioned relation between types of history, he inter alia referred to the intellectual heritage of the discipline he practiced—ethnology (and social anthropology). His vocabulary, as a part of which such terms as 'history' and 'culture' coincide with the term 'race' is symptomatic. This heritage also included the 19th-century trend called evolutionism, representatives of which described the development of peoples in analogy to the development of species analysed by evolutionarily-oriented biologists. In this context, the French structuralist wrote: 'The doctrine of biological evolution admittedly gave sociological evolutionism a decided fillip but the latter actually preceded the former. Without going back to the views which Pascal took over from antiquity, and looking upon humanity as a living being passing through the successive stages of childhood, adolescence and maturity, we may see in the eighteenth century the elaboration of all the basic images which were later to be bandied about—Vico's "spirals", and his "three ages" foreshadowing Comte's "three states", and Condorcet's "stairway"' (Lévi-Strauss 1952: 14–15). Indicating Herbert Spencer and Edward Tylor as the fathers of the (social) evolutionism, the French structuralist pointed out that the theory was 'a pseudo-scientific mask for an old philosophical problem' (Lévi-Strauss 1952: 15). The analogy between the people and the organism turned out to be misleading.[1]

What is interesting is the interest in movement and its dynamics—going back to at least the seventeenth century; at play here is the recreation of the antique thought, as argued by Pascal's example. I am purposefully using the words 'movement' and 'dynamics' next to one another, as the fundamental set of Isaac Newton's laws appeared in the period in question. At the same time, I connect the changeability of movement with an interest in growth. The reference of the social sciences to biology is not accidental. As will soon turn out, biology became the model science in the eighteenth century, and its vocabulary was largely taken over by the subsequently emerging humanities. In this context, the interest in peoples as equivalents of organisms enforced the tackling of the issue of properties of living objects. These, in turn, are not only in motion, but are under development themselves. The etymological meaning of the Polish word for 'development', '*rozwój*', refers to the book in the form of a scroll ('*zwój*'), but in the context of biology and the social sciences, it is valorised in a very special way. However, the problem is how to present this development. I wrote above that it is difficult to express temporal relations in a two-dimensional drawing. When Lévi-Strauss referred to the assumed 'progressive' nature of human transformations, he pointed out: 'As our prehistoric and archaeological knowledge grows, we tend to make increasing use of a spatial scheme of distribution instead of a time scale scheme' (Lévi-Strauss 1952: 21). He suggested that changes should not be presented as a continuous and linear development. Turning attention to the fact that what is at play is more like mutations, 'a series of leaps and bounds' (Lévi-Strauss 1952: 21), he showed an image of the piece of the king in chess and a Yahtzee player, whose progress is open and unpredictable.

In connection with the above, the French ethnologist developed a critical attitude to the image of the spiral, the stages of life and the ladder, each of which testifies to the privileged, progressive nature of movement, the purpose of which expresses the highest and the most perfect state or stage. The tool constructed according to such a scheme was to allow its user to identify not only the general order of the social world, but also their own place in that order. What is more, it testified to temporal relations with the help of basic spatial images. Of course, Lévi-Strauss's critique

---

[1] An example of the medieval Arabian thought is also interesting. In the fourteenth century, Ibn Khaldun, inter alia a pioneer of sociological studies, used the figure of the triad to describe transformations of the Arabian community as a model of human life. Reflecting on the relation between the society and the state, he divided the process of the passage from the nomadic to the settled life into three generations. The first one, represented by Bedouins, was marked by strong group solidarity, which helped the community survive the hardships of life in the desert. The second generation, represented by the established royal authority, which privileged itself at the cost of the group, was tantamount to settling down in, inter alia, the cities conquered during conquest and taking over the forms of life of the local peoples, but also maintaining a relatively strong Bedouin solidarity. Finally, the third stage, represented by the well-established settled life and the monopoly of the king (dynasty), led to decadence, as a part of which the nomadic solidarity was replaced with obedience to the state. The Arabs immersed themselves in the pleasures of life offered by the state, thus becoming cowardly and disregarded by the royal authority (Bielawski 2000: 103–104).

of the thus-valorised and purposeful nature of the development of humanity and its interpretation in the context of 'an old philosophical problem', is understandable, as the order turned out to be illusory. Let us however not forget about the instrumental nature of the tool in question. Here, we are dealing with the readers (listeners) of works, for whom time manifests itself according to a new story—a story seemingly different than the one presented by myths. As a part of this story, time has no end, and its passage, which can be expressed with the help of the figure of the stream or the river, is not limited by any destination. The question is how to depict movement which is not closed, while it seems to be moving forward. Whether or not we are talking about a spiral or states, a road divided into at least three stretches appears. This sort of mechanism of expressing temporal relations with the help of spatial forms turned out to be a relatively—*vide* Lévi-Strauss's words on the antique origin of Pascal's model—original offer.

3.2 The Cycle Revisited

However, in this place, I would like to describe the relation between the cycle, the circle and the spiral with more precision. The mythical story is linear, but as a closed form, it is similar to the drum in that its end is tantamount to a reference to its beginning. We can also find a trace of this operation in science. As shown by the example of research into international communication referring to the figure of U (Kim and Ruben 1988), resolving a cognitive and emotional problem resembles a return to the state prior to a disease. In turn, the spiral is an image suggesting development, and hence can be considered an open form of a cycle, while the circle can be considered its closed form. In this sense, the hermeneutic circle is also a tool enabling progress, although the latter does not reduce to the form criticized by Lévi-Strauss. Simultaneously, we may ask whether development is unidirectional. The already mentioned Vico understood tripartite transformations with reference to the 'corso—ricorso' ('stage'—'return') relation, as it was not just about the forward, but also the backwards movement. Hence, there appears a more refined figure than the antique image of the human transforming, to put it metaphorically, from the puerile to the senile form. However, development is unavoidable also in the case of the Italian philosopher; it is not only guaranteed by the relation between the world and the human—a social and thinking creature—but also by an intervention of Providence. The spiral movement is therefore tantamount to an upward movement from the barbarian to the civilised.

This last relation also emerges in the concept of Vico's contemporary, Henri de Boulainvilliers. What is more, this historian and political writer also used the cyclical approach to time, although he gave it an entirely different form (Foucault 2003). First of all, he replaced the tripartite system with a bipartite one. Rather than as a result of relations between people, the world and Providence, he understood the cycle as a relation of tension between the opposite elements of society, which he referred to as a war. A reference to this relation was a testimony to de Boulainvilliers' awareness that history is a construction rather than a reconstruction of a series of events. In this sense, knowledge—the source of this construction—was of cardinal significance: it was a weapon in the fight for the maintenance of the balance between elements of the social body. Using the subsequent relation originating from the biological vocabulary—'constitution—revolution'—de Boulainvilliers suggested that the state of balance is temporary. In this sense, the lasting of society resembled the movement of the sun in the sky. In the context of writing history, he opposed the royal legal discourse, according to which the present state is the natural state; Boulainvilliers believed that it was a usurpation. The said discourse used the figure of the savage, who through exchange (also the legal one), together with other savages, funded a society, which was to be topped by the ruler. In reality, this story was to obscure the actual state of things, where the ruler was a barbarian conquering the earlier civilisation, which consequently underwent atrophy.

In this place the originality of de Boulainvilliers' operation emerges. His whole project is based on opposites. At the same time, these opposites are understood dynamically. Referring to Ohm's law formulated one hundred years later,[2] we could say that the parties of the opposition remained in unendingly temporary states of balance—a tension which was not changing its value—when the mutual intensity of discourses expressing the relevant knowledge was accompanied by each party's mutual resistance to them. The cycle consisting in the vanishing of a given society was to be a result of an absence of the balance. It turns out, hence, that in contrast to, for example, Vico, de Boulainvilliers did not consider development as a progressing process. Rather than that, he anticipated the abstract model based on the relation of a correspondence between two elements, as noticeable for instance in Durkheim's writings. In this sense, the cycle was not only not limited to the figure of the circle or the spiral, but was secondary to the basic scheme—the relation of opposites. Symptomatically, the relation of war as postulated by de Boulainvilliers is a mirror reflection of the relation of exchange analysed by another representative of the French school of sociology, Marcel Mauss. I consider them both borderline forms of communication (Zaporowski 2018). On the other hand, Vico and de Boulainvilliers used the vocabulary, which a hundred years later founded the concepts criticised by Lévi-Strauss. These concepts seemed to be presenting the individual in the categories of growth rather than a cycle.

3.3 Frazer and Forms of Thought

One of the concepts of social evolutionism was formulated by the British anthropologist James Frazer. His point of reference was the history of thought, which, as can be guessed, he presented in three stages. To this purpose, he used a clever figure of the web composed of threads in three colours symbolising a given form of thought. Initially, the observer noticed the colours white and black, and as the fabric unrolled,

---

[2] $V = IR$, where V is a symbol of voltage, I—current, and R—resistance.

red spots appeared, and the whole became increasingly paler. The author of this tool identified the colour black with magic, red with religion, and white with science. The evolution of human thought was not presented simply as a series replacing a more archaic form with a more modern one. The colour of the fabric simultaneously testified to the progress of thought and the relation between its forms with reference to the type of the society in which these forms dominated. Frazer noticed the similarity between magic and science and differentiated the two from religion. The first two forms allowed an identification of the order of events, although 'the fatal flaw of magic lies not in its general assumption of a sequence of events determined by law, but in its total misconception of the nature of the particular laws which govern that sequence' (Frazer 1993: 49). Hence, although magical thinking erroneously referred to the principle of similarity (instead of the principle of identity), it assumed, just like scientific thinking, some order. Thus, together with scientific thinking, it differed from religious thinking, which assumed a disorder attributed to the divine (higher) intervention.

Frazer therefore perceived a nucleus of modern thought in magical thinking, but nevertheless wanted to suggest a road leading from the erroneous to the correct. The starting point was an improper perception of order. The awareness of error—and Fraser attributed awareness to religious people—led to the negation of order. In turn, the perception of chaos resulted in the renewed, this time correct, ordering of events by scientific thought. This sort of procedure can be compared to the process of the verification of a hypothesis, and indeed Frazer believed that the nature of science was hypothetical. On the other hand, what comes to one's mind is the anticipation of this procedure by the operation of deriving synthesis from Georg Hegel's opposition of 'thesis—antithesis'. Incidentally, I consider this operation a derivative of the taxonomic model based on the relation of opposites and subordination, enriched with the element of time (Zaporowski 2018: 95–96). Anyway, Frazer constructed a tool which allowed to notice the movement of human thought, basically developing linearly rather than cyclically, and at the same time in a single direction rather than back and forth. In this sense, this tool allowed to see progression in the categories of a goal, where the latter 'for ever recedes' (Frazer 1993: 713). Simultaneously, the image of such a road—unpredictable in terms of its results—contrasted in particular with the starting image of Frazer's story, which was mythical.

This is about the myth on the King of the Wood guarding trees in the sacred grove, which refers to the eponymous golden branch. The situation lasted until a daredevil killed him to become a new guardian of the tree with the golden bough, which must not be broken off. Implicitly, the guardian was also a priest and a murderer, while the myth, referring to a pair of Roman deities—Diana and Virbius, equivalents of the Greek deities Artemis and Hippolytus—embodied the unavoidable order of life and death. The similarity of the clearly cyclical procedure of the mythical representation of the world to de Boulainvilliers's historical representation of the relation of war is striking. It was not accidental that Frazer put his model together with the mythical model. He wanted to underline not only the linearity of the story of the advancement of knowledge, but also its open nature. And yet putting together the myth on the King of the Wood and the historical discourse of the French aristocrat seems to indicate both the circular nature of the order of things and the closed character of narration. Do, then, the myth and (de Boulainvilliers's) historical discourse have a common essence? Taking into account the relation of life and death, we may notice its oppositional nature. Thus, an analysis of the myth and the historical discourse allows to see that, in both cases, it is a reference to abstraction—the relation of opposites—which lies behind the circular and closed nature. Therefore, movement does not necessarily testify to time; this is so at least in the case of the closed form—a cycle.[3]

3.4 Purpose and Use

The problem does not lie in the similarity of the schemes of the construction of the scientific and the mythical tool. It lies in the purposes behind these schemes. It should primarily be remembered that the word 'purpose', apart from its other meanings, has the meaning determined by the vocabulary characterising functional relations, including the (teleological) functional explanation. It is not accidental that this vocabulary is universally used in biology, and in consequence also in human science (and, more broadly, in the social sciences). When, therefore, someone asks about the purpose of a myth, then they, first, consider this form of a story as a means, and, secondly, one of potentially many means. Let us assume that the historical discourse is not a myth in the content-related sense and that simultaneously the construction of de Boulainvilliers's tool, just like the construction of the myth on the King of the Wood, testifies to its creator's ability to manipulate (abstract) relations. May the difference between these two forms of a story lie in their placement in different social orders: the one in which divine intervention is acknowledged and the one in which it is not. May now someone point out that both these forms are designed to achieve the same purpose, i.e. evoking the sense of balance, to which the above-discussed category of tension testifies, in the recipient. That person will be right. The use of the functional vocabulary successfully allows to order a very special area—the symbolical sphere or the sphere connected with the worldview—because it is this very area which covers the said stories: the historical discourse and the myth.

However, at least one more element is at play. I am referring here to the achievement of a purpose—which in this case has the form of evoking a relevant reaction. In this place, the myth on the King of the Wood and the discourse on the usurpation of royal power take on a persuasive quality. I wrote above that a given story is a tool with two sides: constructional and representational (or imitative). In another place I also pointed out that this tool is used. In this sense, it is impossible to fully understand the principle of its operation if its performative side, which complements the two other ones, is not taken into consideration.[4] The use—whether we are talking

---

[3] As shown above, not every myth has a cyclical form, although each of them is closed. But the linear myth—which is cyclically recalled—as a part of a custom—fits the scheme in question.

[4] This can be referred to three dimensions of the character: syntactic, semantic and pragmatic.

about the creator or the user of the tool—assumes the presence of attitudes, including propositional ones (Zaporowski 2018). A reference to an (abstract) world of thought is no longer sufficient; it is necessary to refer to socially-shared, although mentally-constructed, ways of assessing these thoughts. What is significant in the situation in question is the circumstance that these ways can be modelled. This is why tools integrating groups, such as myth, testify to the certainly persuasive, if not manipulative, nature of special cognitive acts, including the telling and recognising of stories. The two sides—the author and the recipient—do not have the same pragmatic competence. In this sense, a given interaction does not potentially pose the same challenge to these two parties. This is independent of whether the nature of the story is closed or open.

This is so because also other tools, including the models of social development under discussion, such as Frazer's history of thought, turn out to have a persuasive quality. As much as the myth of the King of the Wood or de Boulainvilliers's discourse allow to understand time not only cyclically but also circularly, and in consequence distance oneself to results of its passage, Frazer's model highlights these results. In its case, the passage of time is to show the progress of the human thought. At least, such a picture is sketched to the recipient of the British anthropologist's story. The recipients should identify the purpose behind the transformations of the mentality of members of the particular societies and place themselves in one of them. In this sense, being fundamentally modern Europeans, they should not deem the purpose reached while being convinced about the privileged position of persons ordering the world from the point of view of science. It is similar in the case of the proposal of Durkheim, who not only positively valorises the modern, but does it, using the functional vocabulary. However, in the context of putting his standing together with Dilthey's position, we should recall that the French sociologist considered the functional relation in the categories of the sufficient and the necessary condition. In Frazer's model, purpose is equivalent to the necessary condition; the relation of correspondence is not taken into account. This seemingly small difference also has serious consequences for the persuasive nature of the tool. Durkheim's relation of correspondence seems to be granting stability, while the relation of looking for the increasingly new (implicitly: better) means for reaching the purpose disturbs this stability.[5]

3.5 Is It Really Progress?

Frazer's model can be compared to another anthropological model developed by a seventy years younger social anthropologist and philosopher—the already mentioned Gellner. Although Gellner referred to the category of the society rather than thought, he adopted the same way of constructing stories as Frazer. The analysed process of

---

[5] One would like to say that a different attitude to the functional vocabulary is a symptom of a difference between the French (continental) apriorism and the British empiricism. At the same time, at play are two forms of the already discussed naturalism.

transformation was tripartite: 'Mankind has passed through three principal stages: (1) hunting/gathering; (2) agrarian society; (3) industrial society' (Gellner 1988: 16). This form of construction of the tool was further underlined by the fact that the process in question covered three basic elements of social practice: production, coercion and cognition. The explicitly expressed system of trinitarianisms was to testify to the complexity of the process of transformation, but at the same time not provide it with an exceedingly unequivocal course. In connection with the above, Gellner concluded that 'There is no obligatory developmental pattern' (Gellner 1988: 16), although, on the other hand, he maintained that transformations could not have a 'spontaneous, endogenous' nature (Gellner 1988: 16). Thus, the British anthropologist rejected the relatively enthusiastic vision of development crowned inter alia by the status of imperial Britain, for which the rebellion of the American colonies was, according to Frazer, the greatest blow. At the same time, while writing about the structure of human history, Gellner presented it just like Frazer—his predecessor and professor of the same school, the University of Cambridge.

It was from Gellner that Kmita took over the concept of social transformations in the context of the drawing apart of the spheres of culture. What is particularly visible in Kmita's thought is his teleological approach to the problem. Of course, according to the Polish scholar—as already mentioned—action is determined doubly, but in every case it is possible to refer to the functional vocabulary as a part of which the purpose is the necessary rather than the sufficient and necessary condition of the means. As results from the above, the vocabulary aims at the presentation of the acting individual in the context of transformations in time, rather than in isolation from the passage of time. In turn, these transformations, although basically having no simple character, are nevertheless presented linearly. On the one hand, such a strategy resulted from the positive valorisation of the modern in comparison with the traditional (or archaic). On the other hand, the linearity assuming a potential achievement of the purpose was aimed at the ordering of movement, i.e. making it more general, thus making the functional relation more than just a temporal consequence of contingent facts. Hence, Gellner and Kmita, being conscious—independently of one another—of the non-obvious and complex nature of the process of social change, continued their way of constructing a tool allowing to approach the human individual—a social animal—in time in the categories of purpose and growth. However, what would the aforementioned non-obviousness and complexity be understood as?

Here, we should mention the second thinker Kmita referred to—Weber.[6] When the Polish philosopher discussed Gellnerian stages, he wrote about types of societies. I am referring here to a special understanding of the word 'type', which was coined by the German philosopher and sociologist in order to refer to an intellectual construct—an ideal type—a generalisation, which simultaneously refers to the value represented by the researcher and is verified by facts independent of them (Weber 2011). This made the construct a product testifying to a high level of self-awareness of the researcher, who controls their valorisation of the object of research. Moreover, Kmita assumed

---

[6] Weber was an anti-naturalist, who, similarly to the British (naturalistically-oriented) empiricists mentioned above, underlined the significance of time.

that a transformation of one type into another takes place owing to what Weber called 'disenchantment of the world' (Weber 1946). Weber used this term to present the process of rationalization. On the one hand, he did not believe, as later picked up by Lévi-Strauss, that members of traditional societies did not act rationally, including—using his terminology—rationally with reference to an end; rationalisation was to consist in the popularization rather than existence of this type of action.[7] On the other hand, without judging whether the disenchantment had a stage-like or a processual nature, the German thinker believed that rationalisation was to lead to the rinsing of the sense from the action, thus creating cynics supporting the annihilation rather than the integration of the society. In this context, the linear progress testified to the paradox of rationality, with the progress turning out to be a victim of itself.

3.6   Demands of the day

Weber's thought was undertaken in one more way. In the same lecture on science, in which the German sociologist and philosopher noticed an ambiguous nature of the disenchantment, he warned against succumbing to objectification. In this sense, he urged to 'set to work and meet "the demands of the day," in human relations as well as in our vocation' (Weber 1946: 156). These words resonated, inter alia, to two anthropologists: Paul Rabinow and Anthony Stavrianakis, who discussed the issue of conducting research, also field research, in changed conditions (Zaporowski 2018). Taking into account the emergence of new areas such as biotechnology and nanotechnology, the Weberian urge turned out to be up-to-date after one hundred years. Generally speaking, the problem the US researchers encountered testified to the progressing rationalisation in the Weberian sense in global science. In turn, the breakdown of the biotechnological project as a part of which Rabinow and his team collaborated with naturalists and engineers, proved to be a special point of reference (Rabinow and Stavrianakis 2013).[8] The collaboration was unsuccessful, since the goals of both parties diverged. While the naturalists and engineers had health amelioration and commercialization of research results in mind, anthropologists focused on the state of human flourishing. This failure however only turned out to be an introduction to further actions.

In this context, a reference to the last sentence of Weber's lecture seems particularly interesting. It is quoted in Rabinow and Stavrianakis's book in the following form: 'This, however, is plain and simple, if each finds and obeys the daemon who holds the fibres of his very life' (Rabinow and Stavrianakis 2013: 4). And yet, the source text has the word 'demon' instead of 'daemon' (Weber 1946: 156), with the former being a translation of the German term 'Dämon'. The question is whether the play on words was accidental. I do not think so. And this was also by no means a

---

[7] Apart from rational actions with reference to the end, Weber distinguished rational actions with reference to value, traditional actions and affective actions. This typology is in principle ahistorical.

[8] Stavrianakis simultaneously participated in a nanotechnological project, which also failed.

purposeful distortion of the meaning. The word 'demon' denotes a demon which in European tradition may be a bad or a good spirit. What is at play in the latter case, as referred to in Weber's words, is a kind of an internal voice inducing people to act. After all, Weber warns against lingering to begin to 'set to work'. In turn, the word 'daemon' refers to a good spirit or deity, and is most often associated with the Greek tradition, including mythology. It is not accidental that the postulated flourishing referred to the Aristotelian understanding of the word 'eu-daemonia', in which the element raising Rabinow and Starvrianakis's interest, i.e. 'Daemon', appeared. Indicating the unsuccessful collaboration as a part of the biotechnological and nanotechnological project, the American anthropologists decided to undertake research in a new configuration, not being discouraged by the previous failure. It would therefore seem that the use of the word 'daemon' was to underline the importance of the goal they had in mind from the very beginning.

However, the case is more complex. On the one hand, Rabinow himself seems to have a rather free approach to quoting Weber's words—in his other book, he used the term 'daimon' (Rabinow 2011: 208), and although it is the Greek equivalent of the Latin form ('daemon'), we could expect a comment referring to the quoted source. On the other hand, Rabinow and Stavrianakis used the words 'demon' and 'daemon' as elements of an opposition, which is to testify to a comparison of, accordingly, the old and the new manner of conducting anthropological inquiry (Rabinow and Stavrianakis 2013: 101–103). What is more, they by no means maintained that one is privileged at the cost of the other. Rather, the clash of the two ways was to result in the emergence of a picture of the real research practice. In this sense, a reference to Weber's words was to indicate that the internal voice or the power directing the actions of anthropologists is complex, thus pointing out to a new context of 'demands of the day', which appeared at the beginning of the twenty-first century. This is because the problem does not lie in the new technologies, but in the new way in which the researcher has to approach themselves. We can see here the impact of Foucault—which Rabinow and his student were by no means hiding. I am referring not only to comparing the individual who steps beyond borders with one who does not think about it, although this thread allows to understand why the American anthropologists did not consider it sufficient to acknowledge that the demon controls their fate.

3.7 The Contemporary

I believe that Rabinow and Stavrianakis's response to Weber's call is a derivative of Rabinow's earlier position in relation to the problem of modernity as formulated by Foucault. Simultaneously, the latter identified the attitude—ethos—and the relation—a relationship between today and yesterday. The American anthropologist, who in the context of his long-term research into biotechnology and synthetic biology as a new field of research at the turn of the twenty-first century, constructed a new figure—the contemporary (Rabinow 2003, 2007; Zaporowski 2018)—also followed

3.7 The Contemporary

this track. It constituted an extension of Foucauldian modernity in the sense that by slightly changing the understanding of the term, Rabinow made the latter an element of the former. From then on, the contemporary—as a relation and an attitude—was to testify to the reflective and critical attitude of the object now referred to as *anthropos* not only to the past, but also to the future. The above was to take place, inter alia, owing to the game of modernity—going towards the new—with the second element of the contemporary—tradition, which was tantamount to going towards the old. What is more, this relation and attitude was mobile in both senses. Firstly, it lasted 'in (non-linear) space' (Rabinow 2007: 2). This sort of conclusion was a testimony to Rabinow's distance to the earlier, classical anthropological tools (and, more broadly speaking, to the ones created as a part of the humanities), which assumed linearity and purposefulness of transformation. The direction of this movement was not only to be unpredictable since; its heretofore nature was to result from a coincidence rather than regularity.

Secondly, the mobile nature of the relation lay in the attitude of modernity to tradition, changing over time. The *Anthropos* was not only no longer a modern or an anti-modern figure. It ceased to be a holistic body, which could be differentiated from other monads. It became a hybrid, the configuration of which was to undergo transformations as time went by. What Rabinow had in mind was not cyclical time. The hybrid covered what he referred to as *bios*, *logos* and *ethos* (the list was open), which themselves mutated; the transformation was most distinctly testified to by the condition of *ethos*, i.e. the said contemporary. However, the vocabulary of the American anthropologist was so flexible that the meaning of the individual terms also mutated in his subsequent publications. For example, 'the contemporary', which became a name for 'a moving ratio' (Rabinow 2007: 2), was referred, in another place—in the context of the relation of sequence—to the word 'the present' (Rabinow and Stavrianakis 2013). In the latter case, this indicated the state or condition of opening to a new way of understanding a research problem, which has not yet been examined as a part of the present. Whereas in the first case of the use of the term 'the contemporary', it was understandable as it suggested a reference of the new to the old, the reference of the term 'the contemporary' to the term 'the present' created a cognitive problem.

No wonder: how to determine a relation between the obvious—the present—and what offers a new solution? In other words, how can we identify a new solution without indicating an earlier solution, but a state which is not a problem, but which calls for a solution? The term 'the present' is, similarly to the term 'modernity', a borrowing from Foucault. The French philosopher who formulated his thought in the categories of opposition, referred today to yesterday. In other words, he explained the present state by earlier events. In turn, Rabinow also opened himself up to the future. This called for a reformulation of the meaning of the borrowed words. Hence, in order to testify to the incommensurability of the positions, we need to refer to another term from Rabinow's vocabulary, which also underwent a specific mutation. I am thinking here about the original term 'the actual', in the context of which 'the contemporary' should be reanalysed. In one of his works, Rabinow used the terms 'the actual' and 'the contemporary' as substitutes for each other not only as a part of the practiced anthropology, but also the formula 'near future and recent past' (Rabinow 2003: 55). Then Rabinow made the formula—reversing the sequence of words—an element of his definition of the contemporary. At the same time, the actual became the object of the contemporary as a part of an anthropology of the contemporary: it was tantamount to a problematization or disturbance of the field of the present (Rabinow 2007). The latter in turn lost its sense as a sole effect of events from the past. Because of the circumstances which later made Rabinow and Stavrianakis clash demon with daemon, the relation between the already three terms had to be established anew.

3.8 The Actual

Hence, Rabinow and Stavrianakis faced the problem of connecting the present with the contemporary, being aware that there is the already mentioned gap between the two elements. They asked another anthropologist, James Faubion, for help and he came up with the idea to use a third element. Let us, however, specify what the two elements indicated. Let me repeat that in contrast to Foucault, the present was not necessarily referred to the past. In this sense, genealogical research was not the focus of explorations. This does not mean that the history of the area was unimportant. Symptomatically, the relation between the past and the present was to be marked by an individual, even contingent causal chain.[9] Nevertheless, the research tool assumed a determination of a starting point marked by a potential absence of problems. This step was conventional. The identification of the situation, which made it possible for the anthropologist to overcome the danger and open up to a new way of approaching it was a part of the same measure—in this place Rabinow and Stavrianakis referred to the figure of the mood. The use of second-order participant observation when the classical participant observation proves unsuccessful, can serve as an example here (Rabinow and Stavrianakis 2013). Does this therefore mean that the latter testified to the third element of the research tool being the object of search?

Not necessarily. Participant observation is a key element of field studies—ethnography. This element is, by the way, a determinant of an anthropological approach to research problems (Reddy 2009). Participant observation lies not so much in the ethnographer's long-term presence close to the problem—the task lies rather in simultaneous co-participation and reflection. Using another vocabulary, it is about the intermittent closing and opening of an electric circuit, and in consequence about the changing of the states of the voltage. The above is to result in knowledge about the changing object, which is updated and immortalised on an on-going basis and from a close perspective. Let us not forget that this object is very tender and sensitive; and it turns out to be, inter alia, the ethnographer themselves. The second-order participant observation—a figure borrowed from Niklas Luhmann (Rabinow and Stavrianakis 2013)—is a procedure as a part of which the ethnographer reflects on their earlier field

---

[9] Rabinow and his team consistently refer to causal relations.

studies. In this sense, the relation between these two types of observation resembles a relation between language and metalanguage. The American anthropologists made ethnography and second-order participant observation components of, accordingly, demon and daemon (Rabinow and Stavrianakis 2013: 102). Hence, since the opening to the new is tantamount to a relational understanding of both these powers, then the domain of one of them cannot precede the state, which it itself constitutes. In other words, ethnography does not precede the second-order participant observation as an element testifying to a disturbance of the obvious, as indicated by the present; rather, it adds to the latter.

Not only was Faubion aware that the relation between the present and the contemporary testifies to the sequence of time. Since the starting point was the state of obviousness, and the point of destination was the state of solving the problem, an intermediate point should be brought into play—the problem itself. It could not, however, be boiled down solely to ethnography. This is because the latter did not lie in dealing with cognitive hazard. When collecting data, the ethnographer proceeds in accordance with a relatively ordered scenario, which is not disturbed by the collected data. The problem could not lie in a change of attitude—the said mood—either, because then it turned out to be bridled. Hence, the disturbance of the order had to be identified. The actual became a proper field of identification as a part of which the classical ethnography proved ineffective, while the shifting of stress to second-order participant observation was tantamount to undertaking a considerable risk of entering an area not yet very well explored. Only a clash of two perspectives—one of them classical, the other new—made it possible to overcome cognitive chaos, while simultaneously looking at the field in a different way. Thus, however, the term 'the actual' acquired a different meaning, since because of the way Faubion used it, it ceased to be a component of a two-element relation expressing spatial relations,[10] becoming a component of a tripartite relation expressing temporal relations. The disturbance—the actual—took place after the state of obviousness and preceded the state of another, although no longer obvious, opening.

3.9 Time

However, it should be stressed that this sort of tool ordering the environment of people as social and cultural animals is non-Hegelian. The heritage of the German philosopher might be indicated not only by the said triad, but also Frazer's earlier idea. It is actually the case in the last example, but Faubion's system does not testify to the purposefulness (or linearity) of transformation of one state into another. It is not accidental, but a conscious decision of subsequent generations of anthropologists that the modernist model of development of societies cannot be maintained.

---

[10] I am referring here to the already mentioned, and postulated by Rabinow, disturbance of the field of the present, where, speaking figuratively, the place rather than the time of action is in focus.

Not only were holistically understood communities (where, interestingly, a community was identified with culture) broken with, but it was assumed that the object of research is tantamount to qualities which not so much last, as emerge. It is against this background that it is worth reflecting on the last case of the construction of the tool enabling us to grasp the process of the transformation of the object of research. It was proposed by Tobias Rees, a German-American anthropologist, who criticised not only the traditional but also his contemporary anthropological vocabulary (Rees 2018). In the context of the classical self-other relation, he concluded that it cannot be analysed in spatial categories. He was by no means referring to a spatial presentation of such a relation, for example in a drawing, as I have already mentioned. What Rees meant was the spatial location of the self, living 'here', and the other, who lives 'there'. In exchange, he proposed that the relation be analysed in temporal categories. He referred to a situation in which the other appeared after the appearance of the self.

This change of perspective was accompanied by a rejection of the vocabulary which contained such obsolete terms as 'culture', 'race' or 'ethnicity'. The human being, as a rational creature, traditionally privileged and contrasted with all other beings, was no longer to remain the object of research. Research was to be focused on thought, or, speaking more precisely, on what has not yet been thought. It is in this context that anthropology, which, stepping beyond its classical form, examined, according to Rees, emergent forms, was subjected to criticism.[11] The focus was on hybrids, which were also to be identified in time. However, the problem was that as compounds, lasting—however briefly—in their constellation, they constituted an object which has already materialised. The object therefore became something, be it potentially, familiar. Dissociating himself also from this form of research, Rees proposed his own version of the anthropology of the actual; this version must have differed from Rabinow's version. The term 'the actual' therefore acquired a new meaning. Rees used it to testify to the on-going passage of time. Since then, the actual was to become an object ceaselessly constituting itself and passing away in front of the observer's eyes or in their self. This allowed the researcher to open up to 'movement as such' (Rees 2018: 105). Although Rees did not indicate it *explicite*, we may say in the context of Hegel to whom he referred that he shifted the accent in his research to becoming.

Leaving the issue of the carrying out of this project in practice open, we should note that the German-American anthropologist remained faithful to the earlier ways of composing the picture of the other. I am referring here to Gellnerian trinitarism. Firstly, Rees referred to the classical form of anthropological research interpreting area spatially. Secondly, he referred to the already time-oriented 'anthropology of the emergent'. Finally, thirdly, he proposed his own version of anthropology of the emergent. Of course, step three has been announced, but from the formal point of view it is as significant a component of the whole as the symbol '0' in the truth-table

---

[11] It is an open question whether the term 'emergent' should be understood as interpreted by a representative of British cultural studies, Raymond Williams, or by the philosophy of the mind and cognitive science. At the same time, it indicates both dynamism, and complexity (hybridity).

(in logic). Neither was it accidental that Rees concentrated on time as a dimension in which the object of research should be interpreted. It is not that the classical anthropology did not appreciate this dimension. This is testified to by the above examples of Frazer and Gellner. It is not so, either, that the spatial dimension was given up later, as testified to by Rabinow's position. Without any doubt, however, it was already Rabinow who parted with the understanding of the process of changes in the categories of purposefulness. It cannot be denied, either, that Rees endeavoured to present the time dimension using the figure of succession. In this sense, movement, even if it did not have any predetermined direction, was not reciprocal or closed. It turned out that Rees belonged to a guild, which manufactured tools enabling the capturing of movement in at least three steps, with the last one not repeating the first one, thus opening itself up to uncertainty.

References

Bielawski J (2000) Ibn Chaldun (Ibn Khaldun). Wiedza Powszechna, Warszawa

Foucault M (2003) Society must be defended (trans: Macey D). Picador, New York

Frazer J (1993) [1890] The golden bough. Wordsworth Editions Limited, Ware

Gellner E (1988) Plough, sword, and book: the structure of human history. The University of Chicago Press, Chicago

Kim YY, Ruben BD (1988) Intercultural transformation. A systems theory. In: Kim YY, Gudykunst WB (eds) Theories in intercultural communication. SAGE Publications, Newbury Park, pp 299–321

Lévi-Strauss C (1952) Race and history. UNESCO, Paris

Rabinow P (2003) Anthropos today: reflections on modern equipment. Princeton University Press, Princeton, Oxford

Rabinow P (2007) Marking time. On the anthropology of the contemporary. Princeton University Press, Princeton, Oxford

Rabinow P (2011) The accompaniment. Assembling the contemporary. The University of Chicago Press, Chicago, London

Rabinow P, Stavrianakis A (2013) Demands of the day. On the logic of anthropological inquiry. The University of Chicago Press, Chicago, London

Reddy DS (2009) Caught! The predicaments of ethnography in collaboration. In: Faubion JD, Marcus GE (eds) Fieldwork is not what it used to be. Learning anthropology's method in a time of transition. Cornell University Press, Ithaca, London, pp 89–112

Rees T (2018) After ethnos. Duke University Press, Durham, London

Weber M (1946) [1919] Science as a vocation. In: Gerth HH, Wright Mills C (eds) From Max Weber: essays in sociology. Oxford University Press, New York, pp 129–156

Weber M (2011) [1904] "Objectivity" in social science and social policy. In: Shils EA, Finch HA (eds) Methodology of social sciences. Transaction Publishers, New Brunswick, London, pp 49–112

Zaporowski A (2018) Action, belief, and community (trans: Moroz-Darska A). Peter Lang GmbH, Berlin

Chapter 4
Change

Abstract The problem of time is associated with my discussion with an American anthropologist T. Rees' searching for time as such. I begin with picking up change as a basic element of movement. Then I take this element not only to be manipulated in representation but also to be a symptom of time. A human action as a sort of event is viewed as implying the change in question. I then author claim that time may be grasped as changes—vide the triad—one confronted with another. A chain of changes is viewed with regard to the end or its absence implied by particular representations. These changes presuppose that a human being does not behave chaotically; the being orders the being's actions with respect to propositional attitudes which are viewed with respect to cognitive acts and cognitive results, where the latter complement one another.

Keywords Time · Movement · Change · Action · Attitude

4.1 Time as Such?

The example of Rees inspires one to ask the question of how time as such can be captured. According to him, does it have an a priori, or an a posteriori nature? In what way can time actualise itself, becoming a research challenge? Since Rees does not provide any answers to these doubts, let us return to the situation when the sense of time is lost—when there is no challenge in the form of time. One example was given by Conrad, when he was describing a journey through the jungle. The protagonist of the novel, Marlow, suspended reflection on stimuli; they were too strong to be controlled. It was similar in the case of the adoption of an aesthetic perspective as described by Geertz. Let me recall that what he had in mind was the human suspending the process of proposition in view of the present stimuli. Finally, the use of the tool, including the drum, allowed distance to time through the reaching of the state of balance. However, the above three cases concern a situation as a part of which people step beyond the daily, common routine marked by the experience of the passage of time. This is not about deliberating on the issue of the motive behind this step, although it is of great significance indeed. In the context of the problem under discussion, it is curious that forgetting—be it purposeful or not—about time assumes its ceaseless recollection. The latter takes place when we identify time not as such, but as a collection of its symptoms. This is what Einstein had in mind when he was referring to the figure of the clock. This is because the question is whether I am able to refer to time without such a tool.

Rees himself was aware that his 'language, the vocabulary' as a tool (Rees 2018: 4) was not perfect enough to solve the problem. This is why, I believe, he devoted so much effort to determining his own position by criticising and distancing himself to the earlier positions. I agree with Rees as to the constructive possibilities of his vocabulary and I consider a reference to time as such as a task which at this point in time is largely breakneck. However, I believe that we may choose yet another road, using the available vocabulary (language), the use of which would allow the identification of symptoms of time and grasping the said dimension through them. This vocabulary would not have to refer to holistic beings of classical (or traditional) anthropology or to emergent forms of non-classical (or modern) anthropology. It should allow current observation of what I call a symptom, which appears in the field of vision or thinking. I use the term 'current', as I am referring to a process or sequence of appearance of increasingly new, unexpected elements. They are interesting as much as they acquire a shape allowing identification. Once identified, they may be rejected for the benefit of a new element. Each of them can be new and surprising, but these properties should be understood relationally. In other words, the uniqueness of something is tantamount to its differentiation from the common and familiar. If then the latter is not taken into account as an element of a tool ordering a given reality, the postulation of the former will become groundless. I do not think therefore that it would be possible to examine the not yet familiar in separation from the familiar. In view of the above, let us assume that we are dealing with an exploration of the relationally-understood current. For the sake of simplification, let us for the time being treat the object of this research as a part of the causal vocabulary.[1]

The said object could inter alia be a process of the appearance of symptoms of time in the field of the experience of the human individual. However, how could this process be reconstructed? It would have to be analysed in the context of the aforementioned control of the researcher over the part and the whole so that they could simultaneously refer the new object to the old one and compare their relation of diversity with the heretofore existing relation of homogeneity covering the familiar preceding the appearance of the new object. One may of course be absorbed by a given object, but we are then returning to the situation of a traveller immersed in exotic and all-embracing surroundings. If, then, a possibility of the researcher's reflection on the said process would come into play, we should assume that it is

---

[1] Let us do it, referring to the problem tackled by Rees inter alia to stick to the convention of modern American anthropologists. At the same time, I wish to point out that when their position is referred to the positions of traditional anthropologists in the context of the relations of 'linear-non-linear' transformations, I shall subsequently discuss this relation as a part of the functional vocabulary. For a discussion of the relation between vocabularies, please refer to my other work (Zaporowski 2018).

possible to separate elements of this process and refer them to each other and to the process as a whole. This is why the symptom must be separated for time to be manifested. When James Joyce experimented with a representation of a stream of consciousness, he showed, be it even accidentally, how—from the technical point of view—this is possible (Joyce 2010). We may imagine a situation when an individual experiences the process of becoming, but does not identify it. Identification takes place when this process is disturbed; when its continuity is broken. We may also talk about a resistance posed by an obstacle to the stream. Let us note that in the last part of his novel, the author of *Ulysses* used a trick consisting in the avoidance of punctuation marks. However, at the same time, he did not avoid either the spacing of words or the division into paragraphs. In this way, the text was not an ideal reflection of a stream—it provided obstacles to the reader, making the intensity of the content divert. He thus, however, allowed the reader to identify the passage of time—its symptoms were not so much words themselves, as disturbances in reading them.

4.2 The Change

The said resistance, disturbance or diversion are no other than changes happening to the experienced stream of stimuli. I am talking about a relation which is worth analysing in particular in the context of the above-mentioned tools allowing manipulation with time. This applies both to the simple mechanism of recreation and complex ways of presentation. This is because the question is whether the change—where sequence is at play[2]—covers one or more disturbances. If more than one disturbed state is at play, is there a relation between them or are they independent of one another? On the other hand, it is not excluded that disturbances should be treated as a discontinuation of a specific state, after which another state will appear. Would this state follow its predecessor directly, or would there be a gap between them? I am talking about a situation to which the principle of continuity (of change) applies (Ajdukiewicz 1978: 203). The Polish logician and philosopher Kazimierz Ajdukiewicz set himself a challenge of facing a certain form of an ancient theorem that change assumes contradiction. He referred to a statement called the principle of transition. It provides that a given body the state of which (A) changes into another state (B), must experience an interim state—different than A and B—during which it transforms from the former to the latter. We should remember that this postulate is a derivative of the aforementioned principle (of continuity), which provides that 'any change must occur in arbitrarily small steps and not in leaps' (Ajdukiewicz 1978: 203).

---

[2] In the methodological sense, this remark is conventional. Let me recall my earlier comment on Lévi-Strauss criticising the principle of causality: as a part of a causal relation, the cause comes earlier than or simultaneously with the effect, but at the same time it is most often a sufficient condition behind the effect.

According to Ajdukiewicz, there are many examples from experience that are consistent with this postulate. For instance, when someone compares the state in which they are at home with their later state when they are in a shop, they also assume the existence of an intermediate state, as a part of which they are walking to the shop. However, let us recall the convention of a comparison of state A with state non-A, i.e. a situation as a part of which the body ceases to be in its heretofore existing state. In consistence with the said postulate (of transition), there must be an intermediate state. If, however, a body is no longer in state A, it is in state non-A. In turn, when the body is not yet in state non-A, in what state is it? It must be in state A. In this moment, the body turns out to be simultaneously in states A and non-A, and this situation is inconsistent with the ontological law of contradiction. Ajdukiewicz perceived a solution in starting with the treatment of non-A as a state in general rather than a particular state. This means that in contrast to the already discussed postulate, non-A can be different from A in any way; the magnitude of the difference in time is not specified. This makes it possible to exclude an interim state which would appear if the difference between A and non-A was defined at any level, while simultaneously maintaining the power of the principle of continuity, assuming change in no matter how small steps. In this sense, the body is in a given state to find itself in a different state within no matter how short a distance of time. What is equally important, Ajdukiewicz pointed out that this principle is an inductive generalisation and that because of that the postulate of transition, being its derivative, is not *necessarily* confirmed by experience. In particular, a conclusion that a difference between two states cannot be of any smallness is not based on experience.

In this way change does not infringe upon the principle of contradiction. At the same time, it turns out that this relation is not necessarily tripartite. After all, one may experience the starting state, the intermediate state, and ultimately the final state. Ajdukiewicz showed that in general, change is a bipartite relation, involving a state which gives way to a subsequent, different state. The tripartite relation turns out to be an overlap of two bipartite relations. In the context of the sequence of time, the second part of the first relation would become the first part of the second relation. And there are two possibilities at play. The example of Frazer, and in some sense also that of Hegel, makes us look at the change as a series of states which are being disturbed. A disturbance seems to be a gap between states, but in the symbolic sense, it is as important as the states with which it neighbours. In this context, Gellner's example as a part of which three alternative, subsequent types are identified, seems to be more obvious. Above all, however, the bipartite nature of change turns out to be fundamental. It is for this reason that the simplest imitative tool such as the shamanic drum allows to comprehend the effectiveness of replication of the basic element of the experienced world and the life it encompasses, i.e. change. I am referring here not so much to the change as such, as to the manner in which the creator of the tool manipulates change.

4.3 Manipulation, Symptom, Action

From among the tools manipulating time, some allow the user (listener, reader) to distance themselves to it, while others underline its presence, or, to be more precise, its property—passage. This does not mean that the former are more archaic or that the latter capture time with more precision. Deprecating the passage of time is characteristic not only for a shamanic drum or a mythical tale (including its entanglements in a custom), but also for a logical scheme referring to the relation of correspondence (a sufficient and a necessary condition). On the other hand, the passage of time does not have to have a particular direction, which suggests that the critical reflection of the subsequent generations of researchers is likely to acknowledge that it is governed by chance rather than order, although at the same time by unavoidable irreversibility. Nevertheless, each of these tools models time, subordinating it to the user's purposes. This happens when—let me repeat—the person steps beyond the daily routine involving the tendency to notice the contingency of time. What is interesting in this context is not only the fact that the said contingency was curbed by constructors of the tools in question in a variety of ways—equally fascinating is the finding of the same interpretation of what a symptom of time is in each of these constructions. The mechanism of the drum lies in the repetition of the same rhythm, while the mechanism of Frazer's narration reflects the alternating and growing introduction of order and disturbance. However what comes into play in both cases is the same element—the change. Is it a symptom of time?

First of all, time is manifested through change. The latter is a relation. Because it is an abstract object, it could be assumed that the grasping of time is solely a domain of thought. However, this is only one side of the medal. It turns out that the aforementioned tools order the change which is experienced as a series of (at least) two subsequent states of the physical world. In this sense, examples of the clock include the repeating patterns of night and day, with reference to which the said thought extracts time from the content of experience. Therefore, the symptom is a (physical) form adopted by the change. Strictly speaking, tools manipulate forms of the change, which means that the change is an element of their mechanism. States of the Earth in relation to the Sun are external in relation to the tool, but the identification of the symptom such as for example the coming of summer after spring—even if this change is relatively hazy—results from the mechanism of the clock. It is because it identifies the change which turns out to be symptomatic. If I now return to the project of examining the current, I shall state that the field of this research takes the form of a collection of symptoms, which appear before the observer at a given moment. Let me therefore repeat that these symptoms are ordered owing to the tool, which understands them as forms of change—a bipartite relation between different states. These forms—symptoms of time—include human action.

Action is a type of event.[3] The latter consists in setting together two states of a given body rather than a persistence of one of them. For instance, if this body is a

---

[3] For the sake of simplification, I am writing about a physical event (and action). At the same time, I am not excluding the presence of a mental event.

stone, an event will not consist in its maintenance of a given position, but in a pattern of two positions. Examples may include a falling or a rolling stone rather than a stone as such or a motionless stone. Events shall not include the state of voltage in an electric cable, but a deviation from the standard such as an interruption or closure of the circuit. Similarly, in the case of action, they will include a deformation, a provision of a new form, rather than an unchanged position. A raised hand is not a gesture—but its raising is; a sound is not an utterance—but a series of sounds may be. An event also assumes a relation between a state and its absence. Examples may include traffic lights or a cyclically beaten instrument. Let us note that rhythm is simply a testimony to a repetition of an event—the same one, implicitly. It is similar in the case of a ritual or custom. Although its message (content) is more complex than a warning against a coming train, the structure is the same. A custom is based on a repetition of the same pattern. It is not accidental that just like the drum, it is used for controlling time. And although time ceases to manifest itself for the people practising the custom (or listening to the sound of the drum), from the constructive point of view, it manifests itself owing to a repetition of the same change in countless symptoms of time rather than putting together various states in various actions.

4.4 Grasping Time

May then an event—and then action as one of its kinds—be a symptom of time. Since, however, time is manifested by change, the latter must be identified in an empirical experience. A contingent testimony will not allow such an operation if the change is not to be contingent or chaotic, so the ordered nature of change can be grasped owing to the ability of abstraction. In keeping with the familiar mechanism of inductive generalisation, which was referred to for instance by Ajdukiewicz, the relation being the object of my interest is separated from the event when the event is repeated several times. Subsequently—reciprocally—as a generalisation, it is used for the ordering of newly encountered events, thus enabling their ordering. On the one hand, the procedure of inductive reasoning is not entirely reliable from the point of view of logic and epistemology. However, on the other hand, already the common experience allows to acknowledge the relative credibility of cognitive results achieved on the basis of this procedure. For example, it is common knowledge that the cycle of sunrises and sunsets is a cognitive illusion; nevertheless, it does not prevent people, equally universally, from basing their daily practices on it. Being a clock used in the Geertzian common perspective, it allows grasping the cyclicity of change, and—through it—the cyclicity of time. It takes place when the observer, having the ability to abstract, can identify the subsequent sunrises and sunsets, i.e. refer them to the previously constructed ordering model.

In connection with the above, change becomes a measure of time. Obviously, change can be assumed by the observer in relation to a single symptom, but only the general nature of an abstracted quality allows to see that time can be teased out in relation to the ordered. First, what comes into play, is the cyclical time which is teased

4.4 Grasping Time

out through the identification of events reproducing the same change. Secondly, time can adopt a linear form when events reproduce two different changes, one after another. However, if change is a measure of time, is the manipulation of change tantamount to a manipulation of time? The example of the cycle of day and night allows to think that what is identified is the order external to the human individual, but the identification—in the form of a cognitive act—is possible owing to a reference to the generalisation of the experience of change, which takes the form of a cognitive result. Owing to the latter, the human individual presents time—conceptually. Deprived of this result, they are unable to identify time—whatever time was supposed to be apart from its symptom. However, the example of the drum indicates a new element. Although its work reproduces the external order—heartbeat—it also enables the modulation of this order. In this sense, the human individual can manipulate change. This does not, however, mean that they manipulate time, which will be testified to by the indications of other clocks—and there are too many of them to be removed from the field of vision.[4] Hence, the said modulation may mean that the user of the drum manipulates only a representation of time. Here, the constructor and the user of the tool are differentiated. The former is potentially aware of the deformation of the relation between the change manifested in action and the time of which it is a measure. The latter does not have to be aware of the above. It is for this reason that the manipulation of change may 'stop' the time for the user for a moment. However, manipulation may also have a different form.

I reflect on manipulation in analytical rather than ethical categories. I write about it not only with reference to the common, but also the scientific perspective. The already mentioned Durkheim and Lévi-Strauss referring to, respectively, the relation of correspondence (function) and the relation of opposition (structure), liberated social solidarity and myth from the problem of change, and, in consequence, time. Regardless of whether it was organic or mechanical solidarity, or the logical or mythical thought which was at play, time was 'immobilised'. The same is true for the position of de Boulainvilliers, which would seem to fit the common perspective, but should be placed in the scientific one. This is because the French historian stepped beyond the horizon of sensory experience and proposed an abstract model referring to an oppositionally-oriented relation of war (Zaporowski 2018). At the same time, the scientific perspective offers yet another solution. Vico, Frazer and Rabinow did not remove time from the field of vision, but provided it with a direction. Whether there was a goal at play—just like in the case of the first two of them—or not—just like in the case of the last one—a direction was tantamount to moving forward in such a way as not to return to the starting point. Therefore, it was possible to adopt a cyclical, but not circular development, or a linear development, or perhaps 'a moving ratio (in non-linear space)'. Let me recall that this was connected with the necessity to distinguish three steps, states or stages. This is because only a triad of elements allows to see movement which does not assume a return. As much as time

---

[4] I am referring here to clocks located potentially near the observer. Above all, it would mean that they are a part of the same inertial frame of reference as the observer.

is sometimes removed both in the common and the scientific perspective, it can also be not disregarded in the latter.

4.5 Perspective and Change

When using Geertz's terminological convention, it is also worth comparing the scientific perspective with the religious one. This is because myth can be located in the latter. I am not analysing it in the context of the common perspective, because as a part of its content, myth, just like the scientific concept, not only steps beyond the world of senses, but also does not have a practical nature; its nature is dogmatic. On the other hand, it would seem that de Boulainvilliers's historical narration is a kind of a myth. The adoption of this suspicion results from the persuasive aspect of this narration, although this aspect—testifying to the pragmatic domain—is also present in scientific narrations. However, in contrast to de Boulainvilliers's narration, myth orders the reality available to the senses through a reference to the ultimate value rather than a construction the effectiveness of which can be subject to verification. Simultaneously, myth turns out to share at least one specific feature with at least one kind of such a construction. What I have in mind is the frequently referred to, closed, and sometimes also circular nature of the tale, which makes time disregarded. Therefore, independently of the motivation behind the narration—be it historical, anthropological or mythical—we may encounter as a part of various perspectives the same scheme of the manipulation of time, which in this case is tantamount to the blurring of its passage (or the significance of this passage); I have already signalled the issue of this passage. At the same time, the term 'passage' is used conventionally here—since I do not maintain that the sensory experience of a flowing stream or river would be the starting point of a reflection on time.

It turns out that the term 'change' is a better word for a measure of time than 'passage'. Ajdukiewicz's example shows, however, that only a deeper analysis allows to liberate the relation of change from the stigma of contradiction. This is so because the change—being abstract—does not easily allow to fit itself into common experience. The latter calls for connecting time with movement, which should be continuous. This is why an image of a flowing river or stream seems to satisfy the condition. However, the observation of animate beings is problematic, since movement seems to be interrupted. Although these beings undergo transformations, there is not sufficient fluidity. It is inter alia for this reason that people notice with a delay that they are getting old or that a new season has come in nature. Anyway, because of the widespread, direct and common experience of various forms of movement, time is associated with—imaginatively rather than conceptually—a passage. Only turning attention to the change of direction or intensity of the stream of experience makes the fluidity become at least an imprecise picture of the situation, if not an illusion. At the same time, this change becomes an object of reflection and an impulse for a distance to the common experience, which makes increasingly new perspectives (as understood by Geertz) emerge as alternative frameworks of the ordering of the world. These frameworks can have a gradually increasing conceptual nature. When, for example, Hilary Putnam differentiated extension from stereotype, he pointed out that the former does not have an imaginary character (Putnam 1975).

If the change itself becomes an object of reflection, we will be able—as indicated above—to see a testimony to time in the event which assumes it. On the other hand, this reflection will enable, inter alia, a possibility to manipulate the representation of time so as to curb the experienced unpredictability of the change—also as indicated above. However, as it turns out, it is not true that each of these perspectives is responsible for a separate manipulative strategy. For this reason—and I am referring here to a comparison of de Boulainvilliers's narration to a myth—what comes into play is a division which runs crosswise to these perspectives. I am talking about an approach which disregards change or makes it an element of a balanced system, as well as an approach underlining the significance of change as a testimony to the imbalance of the system. This is in particular visible as a part of the scientific perspective, which encompasses the human as a system which does not act on an impulse. The social and cultural dimension of action turns out to allow the adoption of any of these two approaches. At the same time, the change does not cease to be a troublesome object of analysis. On the one hand, if the change is disregarded, the researcher will live in the world of immobile and uniform monads—cultures and societies—the persistence of which is not confirmed by experience. On the other hand, if the significance of change is highlighted, the researcher will face the consequences of a transformation of one—cultural and social—form into another, potentially unpredictable. I am referring here to both the effect and the purpose of the said transformation; I am also certainly talking about its necessary condition. Simultaneously, this is not about the forms in question as such, but about their detailed research constructions, a couple of which have been presented above.

4.6 Absence and Goal (Effect) of the Change

Hence, in the context of the research on the current, as reflected on above, change should not be disregarded simultaneously with privileging the goal (or effect) of the events which assume it. I am referring here to the change which takes place in the domain of culture and society, where these two constructs are different sides of the same coin. If I refer to the figure of the humanities, I shall have in mind the perspective as a part of which (at least) two determinants of human actions are identified: one of them will be responsible for the ability to order them (culture), while the other one—for the ability to share this order (society). I have suggested the applicable vocabulary allowing the mutual reference between the said order and its sharing in my earlier work (Zaporowski2018), and now I would like to use it to refer to the above statement. As a starting point, let us therefore use the thesis that a physical event assuming change is an ineffaceable element of experience. Speaking in more detail, experience becomes significant upon the identification of change, which can be grasped as a part of an event such as a human non-impulsive action. At the same time, such actions turn out to be such cognitive acts as I discussed above. This means that their elements include the taking of an appropriate propositional attitude such as a belief. In other words, the adoption of a belief is interwoven with a change taking place in the cognitive act—action—with the change being—let me repeat—a measure of time.

I would like to turn attention to the fact that since the cognitive act is an element of the physical world, the change taking place as its part is a change of the world.[5] Simultaneously, this act is a reaction—since the acting individual enters the network of interactions with the rest of the world—to an event such as an action of another individual. In view of the above, the identification of change taking place as a part of a given cognitive act is tantamount to the experience of changes taking place in the surroundings of the author of this act. For instance, recalling during a walk that we have already seen the same landscape is as cognitively significant as bowing on seeing an acquaintance. In both cases the change lies in the shifting from the passive to the active state; the landscape and the person undergo identification rather than are identified. This change is a consequence of an earlier change, i.e. respectively, finding ourselves in a certain place and meeting someone. In turn, the cognitive act in question will potentially cause subsequent changes and in consequence will co-create a network of events, including actions, which, being symptoms of time, will allow grasping the world as a mobile system. However, this (i.e. the act of grasping) does not necessarily have to be so. This is because not only the landscape, but also the encountered individual do not have to refer to time. Things are different, however, with the author of the cognitive act. Even if in a given situation the individual does not have to be aware of the act which is taking place, then it is possible *post factum*. Examples might include the already mentioned ageing or the sequence of seasons. In this context, it is not recommended to disregard change, in particular when it concerns something more than common experience.

If this individual steps beyond this experience, they are willing to perceive the change as necessary. This is what both the creator and the recipient of the myth did. In the case of the two, different motives might have come into play. For example, the former could operate as a part of a religious perspective, and hence manipulate ultimate values of the public, while the latter might not want, but have to succumb to this manipulation. Anyway, a goal accompanied a series of actions—writing, telling or reading of a myth—or an effect was analysed. Also in this case these actions were cognitive acts, because they made a framework for the experienced world. At the same time, such acts include actions testifying to concepts explaining, inter alia, persuasive narrations encountered in the humanities—after all, the pragmatic domain is present not only in the religious perspective. One group of these narrations is composed of the ones which connect the effect (goal) with the unchangeable and abstract. In their context, the experienced change—in the constructive sense—makes up a universal mechanism providing the user of the given content with the sense of balance. Concepts as a part of which the change is attributed to the goal (effect), i.e.

---

[5] Obviously– let me repeat my earlier remark—change is understood through a reference to the definition of the term 'change' in the research vocabulary.

the necessary direction of movement which makes an element of such change, appear in the second group. In this case, the change has as an equally significant constructive nature and the difference lies in making the recipients of such a concept aware that they are entangled in the process of transformations, which potentially exposes them to an unpredictable future. As much then as we should view the concept, which just like myth provides us with the sense of balance, critically, it is worth to distance ourselves to the concept continuing the mythical tale of destination.

4.7 Absence of the Goal (Effect) and the All-Embodying Nature of Change

In turn, what would be the consequences of the researcher's distancing themselves to the goal (effect) of the cultural and social change in relation to the domain of actions of an individual? In this context, Rees takes the most radical position (Rees 2018). He not only rejects the idea of the goal (effect), but also is determined to grasp time directly, regardless of any testimony. Therefore, going back to my earlier reflections, let me ask whether changes can at all be at play? Of course, if the anthropologist intends to tackle what has not yet been thought, rejecting the already familiar forms, the change would be tantamount to at least a discontinuation of the earlier research practice. However, the above excludes change as the object of analysis, since the object is yet to emerge; it is not known at the present moment. The option Rees calls the anthropology of the emergent seems to be more promising. Let me recall that its supporters assume the examination of mobile and contingent hybrid forms which undergo transformations into new forms in new configurations of their elements. In this case, the object manifests itself processually, and although it melts away, it does allow brief identification. Could not we then diagnose change, which, deprived of any foreseeable effect (goal), nevertheless progresses in a direction? This direction could be traced by not only looking for the possible forms subsequent to the already experienced form, but also by reconstructing the earlier forms. This practice would at least allow to search for new ways of opening up to the future—as a part of the contemporary—owing to an analysis of the problematisation of the present in the field of the actual.

The problem lies in grasping the form the change takes. As already mentioned, the emergent is a hybrid. The latter is not only a temporary, but also a compound form. Parts of this compound are potentially mutually 'unfitting'. On the one hand, each part has its own history or trajectory, as it needs to be understood dynamically, in motion. The misfitting of the parts means that the whole does not match a familiar scheme; it also means that the whole is temporary, since the mechanism of dovetailing one part with the other ones, allowing the maintenance of relative permanence, is not known. On the other hand, as much as a given part is to have its own history or trajectory, its uniqueness remains unclear. This is because the question is whether this part is a holistic (coherent) construct or whether it is a hybrid itself.[6] If the former is the case, what principle allows one to assume that the pure adjoins the mixed? This problem was tackled by Ludwig Wittgenstein, when as a part of his logical atomism, he postulated that an elementary proposition is a truth function of itself (Wittgenstein 2001), which leads to the identification of the whole with its part. If the latter comes into play, a question arises whether the supporters of the hybrid understanding of the reality allow disassembling of a given whole into parts, which, being compounds themselves, will enforce a never-ending analysis of their structure. In connection with the above, do they also control the relation between the whole and its part? Hence, the hybrid turns out to be a perhaps impressive analogy of constructs of synthetic biology such as explored by Rabinow. Nevertheless, as an analogy per se, it is not necessarily an appropriate testimony of the occurring transformations outside the area of biotechnology.

It is for this reason that I have decided to use the event as the said testimony. It is naturally a research construct; it is conventional. However, it allows to understand the dynamic and unpredictable object in a potentially cohesive manner. Being the smallest unit of analysis, it is itself a holistic construct. If the current is to be diagnosed in the field of experience, then its significant nature will be reconstructed as a change taking place as a part of an event. Let me repeat that an absence of change does not introduce new information about the process under study. Moreover, an event testifying to change may co-create larger wholes. What is at play in this context is not only time, which will be thus teased out from the testimony. When I previously wrote about the human being, I understood them as a bundle of actions (Zaporowski 2018). The figure of the bundle suggests that not only time, but also space is a point of reference for an analysis of change. Speaking more precisely, what comes into play is space–time as a part of which the individual as a bundle of actions participates in many simultaneous interactions with events of the physical world, including other people's actions. This makes the world become a bundle of events. The aspect of, inter alia, the spatial location of a given object was noticed much earlier, as testified to by Rees's critical analysis of classical anthropology on the one hand, and Rabinow's definition of contemporaneity on the other. I am referring to this aspect of the issue for ordering purposes, aware that these considerations are mainly focused on one of the elements of the four-dimensional whole testified to by the relationally-understood physical event.

---

[6] I have discussed this problem earlier (Zaporowski 2018).

4.8 Order

When I assume that an event has a relational nature, I refer to the instance which orders it. It is owing to this instance that it ceases to be contingent. I have already mentioned that elements of a cognitive act—a very special event[7]—include the adoption of a position. This inter alia means that I understand the cognitive act pragmatically. On the one hand, the element is not experienced—perceived—by the observer. The latter can only presume that their partner to the interaction adopted an attitude on the basis of his/her behaviour and in this way reconstruct the mental background of the acting individual. The observer will do so, as they will ascribe to their partner the attribute—the possibility to adopt an attitude—which they ascribe to themselves. This action will not be tantamount to calling into life any (mental) beings; I understand this procedure in semantic rather than ontological categories. The ascribing of the said attribute will be tantamount to referring to special research vocabulary as a part of which an image of the individual's actions will be constructed—the actions being the movement of the body which makes mistakes, corrects them and chooses—in advance—a new direction, rather than the movement of a machine—a *zombie* or artificial intelligence. In this sense, the word 'mental' does not suggest anything beyond a conclusion that cognitive acts become ordered, when they are not only generalised, which is possible as a consequence of testifying to the abovementioned predicate of satisfaction—what also comes into play is the individual's ability of a relatively constant assessment of cognitive results formulated through acts.

Thus, the adoption of an attitude assumes a reference of a cognitive act to an attitude which is neither action nor any of its components. Instead, it is a way of making an assessment of the said results. The term 'attitude' is tantamount to an object postulated by the aforementioned research vocabulary, which constitutes a mental reality.[8] When I used this term previously, I had in mind propositional attitudes, i.e. attitudes adopted to what analytic philosophy, the philosophy of the mind and cognitive science considered the meaning of a sentence, i.e. a thought or a proposition. Attitudes are considered an effect of an individual's interactions with the world and are responsible for a relatively stable and foreseeable character of human actions. First, they can be understood habitually, as a generalised ability to respond to events, while secondly, they can be ascribed the quality of human calculations and designing innovative steps. In both cases, attitudes can be adopted while acting, respectively, habitually or creatively. Differentiated from the (said) *qualia*, they are not private; although they are attributed to individuals, they can be shared

---

[7] What comes into play is two events, or, to be more precise, one event available under two labels: a physical event and a mental event. The former may be exemplified by an utterance of a sentence, while the latter—by recalling someone's utterance.

[8] It is an archaic vocabulary, but useful enough to allow the placement of mental events in the causal chain together with physical events or putting both event descriptions together as a part of a functional relation. In this sense, inter alia following Donald Davidson's anomalous monism (Davidson 2001), I opt against the reduction of the mental to the physical.

socially. Beliefs, intentions and desires—to mention but a few types of attitudes—allow people, by supporting one another, to connect the entire scope of their actions with other events. The Foucauldian contemporaneity is also an attitude, although it has been introduced to the game as a part of a slightly different research vocabulary. It did not concern meanings, but the present, i.e., directly, events requiring a critical distance to, among other things, oneself. It turns out, however, that the attitude orders human actions also in this case. Since the changing world of events is—potentially—able to enforce a change of a configuration of attitudes, the whole created by them is marked by flexibility. This allows the individual not only to adjust themselves to the changing conditions, but also constitute them.

The simultaneously ordering and permanent nature of attitudes is significant. Also, not being events, they do not assume any change. They compound the reaction of the self to the appearance of abstract objects—notions—which, understood relationally, create the aforementioned meanings of sentences. These meanings have been referred to above as cognitive results. When interacting with the world, the individual performs many cognitive acts, and this experience results, inter alia, in generalised cognitive results, while simultaneously, in ordered ways of their evaluation. Let us imagine that the latter obtain a particular form, becoming (affirmative) propositions, which from then on may be considered elements of tools such as myth or scientific concepts. The significance of attitudes lies in the fact that without their participation, the usefulness of the said tools becomes troublesome. More specifically, what becomes problematic is the identification of the use of the cognitive result adopted by its author. This is why an analysis of an unfamiliar text, deprived of the pragmatic element, does not bring us any closer to either the purposefulness of the determination of the actions of the protagonist of a myth or the justification behind the identification of the effect of the event under explanation. Cognitive results may become immortalised, although they do not have to be recognised appropriately—or at all. This is because the attitudes are not only generalised ways of coping with cognitive results—they are an effect of the spatiotemporally defined interaction of the individual with the world, so that the individual knows the scope of the applicability of the results in particular cases in the space–time that individual is a part of. If then the interpreter has no access to the experience of the constructor of the given tool, then their attempt at the reproduction of the configuration of attitudes of that author is much more difficult. Of course, this does not mean that the interpreter shall not use the said tool, but they will have to take into account the problem of Wittgenstein's private language (Wittgenstein 1958).

4.9 The Act and the Result

The cognitive act and the cognitive result are separate qualities. The former is specific, while the latter—abstract. What is more, the former testifies to the process, while the latter—to structure. Nevertheless, they complement each other. This is testified to by the medium located at the intersection of the two domains co-created by these

4.9 The Act and the Result

qualities. This medium has the form of a sentence belonging to a given language, which refers—obviously in a different way—to both the act and the result. On the one hand, the sentence can be used—for example uttered. In this case we are dealing with a cognitive act. On the other hand, the sentence has a meaning taking the form of a proposition.[9] The latter is a cognitive result. The problem we do not perceive when trying to harmonise the domain of the act and the result is the pragmatic nature of the former, which the latter can do without. After all, language as a syntactic-semantic structure does not require a carrier. In this sense, even if—by referring to the category of the sentence—we may establish a relation between an empirical and a theoretical sentence (where the former to a certain degree testifies to the cognitive act, while the latter refers to the cognitive result), this relation shall not take into account the human factor. Obviously, both the act and the result are attributes of the cognizing individual, but only the said pragmatic dimension of the sign allows to see that without being subjected to interpretation or the attribution of, inter alia, attitudes, the cognitive result becomes unrecognisable for the individual, as has been pointed our earlier. In this sense, the act becomes a specific testimony to the human agency.

The complementary nature of the cognitive act and result may be pictured by Foucault's and Kmita's positions quoted at the beginning of my deliberations. On the one hand, a change is perceived in the act of the adoption of an attitude of modernity. The non-modern individual, who succumbs to the passage of time and potentially does not reflect on their own condition, treats the world as continuous. The breaking of the continuity becomes possible as much as the said attitude constitutes the human self. Of course, it is not so that modernity is present as such. It emerges similarly to propositional attitudes—on the basis of experience.[10] It is not settled either that in the context of other attitudes its specificity may change. Anyway, Foucault pointed to the performative nature of being modern, which lies in ceaseless—processually understood—attempts at stepping beyond the borders of one's own, heretofore existing, condition. Hence, it can be said that a series of cognitive acts as a part of which change is identified results in new cognitive results testifying to the condition. On the other hand, change is noticed when various propositions being elements of both spheres of culture postulated by Kmita are compared. As shown above, this is not about the cognitive results these propositions are, but their interpretation, i.e. the involvement of one's own attitudes in the assessment of these propositions. I consider the period as a part of which, according to Kmita, the differentiation of these two spheres took place, particularly significant.

When writing about this period, I refer to a specific spatiotemporal system—may it be ancient Greece—as a part of which a series of changes taking place among the

---

[9] Of course, the problem of recognising a proposition as the meaning of a sentence may seem no longer applicable. I am referring here to Ajdukiewicz's position. However, I treat it in semantic categories—I am not going to simultaneously entangle myself in metaphysical disputes.

[10] May it, for example, apply to the experience of the consequences of non-critical obedience to the Kantian guardian.

religiously-oriented communities, consisting in the gradual differentiation of integrative actions from instrumental ones, was noticed. I consider it significant that the said differentiation was tantamount to respecting increasingly new patterns of propositions different to the previous ones, which called, at least from some individuals, for being modern as understood by Foucault. Anyway, the taking of subsequent actions—cognitive acts—indicated the ability to identify change, with it gradually being taken into account in the content of the cognitive results. For example, if Kmita points to the special role of normative propositions, i.e. propositions identifying the purpose behind the actions of members of a given community, then this goal could be extracted from Foucault's narration—a series of propositions. It is the need to be modern. In this way, both thinkers allow us to see that the exploration of the current is not focused on the examination of time as such, but its symptoms, which are ordered from the point of view of the relation between the cognitive act and the cognitive result. These symptoms, understood as events testifying to change, are related to the individual whose self is co-created by temporary systems of attitudes. I have already called a very special form adopted by change a crack. Further on, I shall elaborate on what I had in mind.

References

Ajdukiewicz K (1978) [1948] Change and contradiction (trans: Giedymin J). In: Ajdukiewicz K (ed) The scientific world-perspective and other essays, 1931–1963. D. Reidel Pub. Co., Dordrecht, pp 192–208
Davidson D (2001) Mental events. In: Davidson D (ed) Essays on actions and events. Oxford University Press, Oxford, pp 207–225
Joyce J (2010) [1922] Ulysses. Wordsworth Classics, Knoxville
Putnam H (1975) The meaning of "meaning". In: Putnam H (ed) Mind, language and reality. Philosophical papers 2. Cambridge University Press, Cambridge, pp 215–271
Rees T (2018) After ethnos. Duke University Press, Durham, London
Wittgenstein L (1958) [1953] Philosophical investigations (trans: Anscombe GEM). Basil Blackwell, Oxford
Wittgenstein L (2001) [1922] Tractatus logico-philosophicus (trans: Pears DF, McGuinness BF), Routledge, London, New York
Zaporowski A (2018) Action, belief, and community (trans: Moroz-Darska A). Peter Lang GmbH, Berlin

Chapter 5
Crack

Abstract This begins with recalling three artistic works where the crack in question is recognized as a sign of irreversible change. These works are then related to the possibly dramatic experiences of both common people and scholars. These are cultural shock, stultitia and falsification. I then show the ways of dealing with these experiences, and propose to consider three scenarios. The first is that it is a community where, for example, myths are the remedy. The second is that modern individuals, being modern and therefore self-critical, can find their remedy. Finally, the third scenario suggest it is a couple—represented by real scholars—who mutually reinforce each other to free themselves from trauma.

Keywords Crack · Cultural shock · Stultitia · Falsification · Remedy

5.1 Mandelstam

You brute of a century, who could look

into the centres of your eyes

and with their blood glue back

two centuries to a severed spine? (Mandelstam 2021)

With this question, Osip Mandelstam, a Russian poet of Jewish origin, begins his poem *Century* dating to 1922. The author refers not so much to a century as to an epoch, pictured as an animal, which used to be 'lithe', but is inert today, having just 'the trail of its own claws' in front of it (Mandelstam 2021). Its grandness has passed away, since something was interrupted. In the quoted question, the formula 'a severed spine' indicates two vertebrae, which were disconnected. On the one hand, in some languages, including Polish, the figure of the vertebra (*kręg*) brings to mind a cycle, including a circle (*krąg*), suggesting cohesion and harmony. On the other, the figure testifies to a part referred to a whole—the spine. In Mandelstam's poem, the backbone stands behind the vitality of the epoch, so talking to the latter, which takes the form of the beast, he writes: 'your spine's now thoroughly smashed' (Mandelstam 2021). The dynamic and creative epoch is presented as a body whose life goes in a circle. The circle, represented by the vertebra, i.e. a part of the spine, is an element of the whole, which is imperceivable. This means that the moving body—the developing epoch—is deprived of consciousness. The whole is grasped only when the spine is broken. But this is a testimony to reflection on something which has passed away. Hence, the consciousness of an event is tantamount to its irreversibility. Parting with the old epoch is bitter, as the reaction to the drama takes the form of the 'indifference' of the new epoch. The latter pulsates with life, develops and undergoes the same cycle of development until its 'smashed spine' enables the appearance of a subsequent 'soft gristle' (Mandelstam 2021).

Mandelstam's poem was recalled by Rabinow (Rabinow 2011). He did so, referring to the Italian philosopher Giorgio Agambena's position concerning contemporaneity, which—let me recall—he also defined himself in the context of his anthropology (Rabinow 2007). I have already compared the two positions (Zaporowski 2014), and now I only wish to signal what connects them in the context of the interpretation of *Century*. Analysing contemporaneity, Agamben and Rabinow stressed the importance of Friedrich Nietzsche's idea of the untimely, which suggested distance to the present. This idea fits Foucault's later idea of being modern, which was clearly analysed by at least Rabinow. Agamben's understanding of contemporaneity assumed a reference to the past founding the present, including the Heideggerian figure of the *arché*, while Rabinow understood it relationally—as a purposeless movement between the past and the future. Thus, the former stressed the primeval nucleus in the present, while the latter did not envisage the determination of the present by the past. In this sense, they understood contemporaneity in a different way. They both, however, understood contemporaneity as a treasury of reflection on the present. In this context, *Century* is a very special cognitive result. This result does not have a generalising nature, like Ajdukiewicz's scientific proposition. It has a principally aesthetic value consisting in a shocking presentation of the interruption of the continuity of the surrounding world. This is so because this reflection is not a result of just any change, but the collapse of the whole world.

For this reason I believe that it is worthwhile to analyse this cognitive result deeper, stepping beyond Agamben's interpretation, which was critically referred to by Rabinow. I am referring here to the experience of an individual, who lived not only at the turn of two centuries: the 19th and the 20th. This is what is discussed in the quoted fragment of the poem. Mandelstam was born in the western edge of tsarist Russia and died in the easternmost end of Soviet Russia. He was born as a Jew in multi-ethnic Warsaw, left this world as a prisoner of a gulag, and was buried in a collective grave. The said two centuries then are also tantamount to two different worlds severed by the experience of the First World War and the two revolutions of 1917. When Mandelstam wrote about the old age which is to give way to the new era, he was aware of the irreversibility of the fate of not only the country he lived in, but also a certain social and cultural formation typical for the nineteenth century as portrayed inter alia by Proust (Proust 2003). The title of the French writer's novel—*In Search of Lost Time*—is meaningful in this context. However, Mandelstam experienced a special trauma, since he witnessed the approach of de Boulainvilliers' barbarian and the murder of Frazer's King of the Wood. Hence, the figure of the vertebra he referred to does not have a positive nature such as it had, for example, in Vico's writings. Here, the cycle of life does not end with a natural death, but a disaster—which is heard rather than seen because here one can hear a crack. Above all, however, it is a cycle which comes into play. In this sense, *Century*, being a cognitive result, also becomes a tool reflecting time marked with a markedly archaic provenance.

5.2 Nabokov

For me, Vladimir Nabokov's novel *Bend Sinister* is a similar tool. The Russian writer and immigrant published it in English in 1947. It is devoted to the fate of an individual subjected to the terror of a totalitarian state; at the same time, it is the writer's specific game with the reader. The main protagonist, the philosopher Adam Krug, loses his wife, and then his son, being under a ceaseless pressure of his former school mate turned dictator—Paduk. In the parallel world, indicated at the beginning and end of the novel by a puddle on the road—a ceaselessly renewed mirror and passage—the author, pointing to himself, liberates the protagonist by falling into madness. The novel is also a specific linguistic labyrinth, as Nabokov not only referred to many earlier works of international literature, but also applied his own experience. Here, for example, in an imagined world, people speak a mixture of Russian and German. Nabokov, born in Russia after Bolsheviks rose to power, escaped to Germany, from where he also had to flee in 1937, when the Nazi terror flourished. Although, therefore, the author himself did not explicitly indicate the connections between the totalitarian system he became acquainted with and the problems of the novel, its content makes one believe that the fictitious world directly referred to the 20th-century reality. *Bend Sinister* was by no means an exception in Nabokov's output from the point of view of the presentation of relations between the individual and a dictatorship, which made him develop a personal and emotional attitude to the protagonists of his novels.

The said linguistic labyrinth can be exemplified by a play with the terms used in the novel. The surname of the main protagonist is one of them. According to Leona Toker, the word '*krug*' is tantamount to both a 'vertebra' and a 'circle' in Russian,[1] and a 'pitcher' in German (Toker 1989: 179). The former would suggest the world presented by Mandelstam, but the latter refers to the pitcher's 'curvature of (…) handle' (Toker 1989: 179), which in turn might be associated with the title of the novel. The term 'bend sinister' has a heraldic meaning. It refers to a stripe, which runs down diagonally to the left on the escutcheon. However, Nabokov himself wrote in the Preface that he wanted 'to suggest an outline broken by refraction, a distortion in the mirror of being, a wrong turn taken by life, a sinistral and sinister world' (Nabokov 1974: 5). The protagonist's surname would therefore suggest a bend or

---

[1] In Russian, the word *krug* means both a vertebra and a circle.

interruption of the course of life, which actually happens in the novel. Should we therefore be guided by the figure of the straight line, which is broken, like a ray of light in a prism? We might think so. However, on the other hand, Nabokov continues: 'The main theme of *Bend Sinister* (…) is the beating of Krug's loving heart, the torture an intense tenderness is subjected to (…)' (Nabokov 1974: 7), referring to the father's bond with his son, David. The aforementioned Toker indicates that the novel includes many references to the figure of pulsation (Toker 1989: 180–181). She also quotes a fragment in which Paduk's heartbeat can be heard, accompanied by 'an additional systole, causing a slight break in the rhythm' (Nabokov 1974: 123). According to Toker, the interruption of the rhythm testifies to the fate of Krug rather than Paduk.

If the reader of the novel goes back a couple of verses before the quoted fragment, they will read that Krug can hear 'a steady thumping sound not unlike that of an African drum (…)' (Nabokov 1974: 122). May they now compare the quotation with the quotation concerning the main theme of the novel. First, they will notice that, perhaps subconsciously, Nabokov referred to Conrad. The connection between the heartbeat and the sound of the African drum is striking. Secondly, the reader may think that the figure of the bend may be circular rather than linear. The said pitcher also has a podgy shape. The circularity may be manifested, just like in Conrad's text, in the rhythm in which the drum is beaten, which imitates the rhythm of the beating heart. Thirdly, Nabokov's image differs from Conrad's image by an indication of an interruption of the rhythm. This is because *Bend Sinister* is about the highlighting of an interruption rather than the identification of order. The motif of Nabokov's novel is not to be a repetition of Conrad's motif, but its transformation. Hence, fourthly, *Bend Sinister* can be interpreted as a tool, which simultaneously uses an earlier mechanism—although there are many more references to literature in this book—of presenting time, and marks change in a novel way. The latter takes the form of a disruption of the order, which is not a symptom of a disaster, as in Mandelstam's work, but of degeneration—one tantamount to succumbing to the dictate of violence or becoming mad, and finally to a violent although unconscious death. The change understood in this way will not be accompanied by indifference, as it will turn out to be too sinister.

5.3 Cohen

Ring the bells that still can ring
  Forget your perfect offering
  There is a crack a crack in everything
  That's how the light gets in. (Cohen 2021)

This is a refrain of the song *Anthem* from Leonard Cohen's album *The Future*, which was released in 1992. The lyrics paint a pessimistic picture of the world contemporary to the Canadian-American singer. Cohen sang about everlasting war, captivity, lawlessness and lies. The objects of his concern also included the imbalance and the insecurity in the intellectual and emotional sense. It would therefore seem that there is no help for life in such world. This suspicion is strengthened by the eponymous song, in which murder is one of the signs of what is approaching. However, Cohen did not intend to leave his listeners believing in the inevitability and victory of evil. The words of the refrain can be interpreted as a call for action. The author seems to be saying that while assuming that we live in an imperfect world, we should act appropriately here and now, at the possibly individual level, rather than refrain from hiding ourselves behind lofty ideas. The fact that things are cracked does not mean that they are worthless. They require a repair, even as imperfect as they are. Cohen assumes that we are such things ourselves. Until we remain whole, we are not aware of our condition. It is only through cracks that the light of knowledge on ourselves, on us being unavoidably defective, can reach us. The pessimism of the initial image is therefore contrasted with an optimistic message that defectiveness is a challenge rather than a curse.

Cohen's song does not end as dramatically as Mandelstam's and Nabokov's works. Perhaps this should be attributed to the fact that the fate of the author of *Anthem* was not as traumatic. What connects him with them, however, is the performance of a similar dramaturgical trick. It is about turning attention to the rapid interruption of continuity as an effect of a poignant experience of the individual in the restless world. In every case, the event takes on a different form. Nevertheless, it is interesting that Cohen's figure of the crack fits Mandelstam's figure of the crack of the breaking spine and Nabokov's disturbed heartbeat, the cracked pitcher, and even its breaking. In the latter case, I am referring to the German meaning of the word 'krug' as well as the scene in which the bullet shatters the protagonist's head. I therefore want to underline a similar sensitivity of the three authors, who—although going in different directions—provided a similar diagnosis of the human individual condition of being vulnerable to a cognitive shock and an impossibility to return to the state prior to the disaster. In this sense, in contrast to many myths, which may resemble the works under analysis, time is strongly highlighted; its passage turns out to be irreversible. On the other hand, Cohen seems to be less of a pessimist than Mandelstam and Nabokov. A broken spine cannot be glued together, a scratch on a pitcher cannot be removed, but all this may become a condition for rebirthing rather than remain a mark of destiny.

Cohen's ethical message becomes the more intense, the more the reader starts to thicken the meaning of the quoted fragment of *Anthem*, in particular the motif of crack and light. I myself have found at least two interpretative paths. The first one shows Cohen as an Ashkenazi Jew who refers to kabbalistic tradition, including Isaac Luria. This descendant of Sephardi immigrants from the Iberian Peninsula, born in Palestine, was the author of a system explaining, inter alia, the reason behind the origin of the material world—the evil world. This world was created from remnants of shattered and fallen vessels, in which divine light was kept, which had been largely rescued from the fall. However, some sparks of the light were trapped among the remnants, so the goal of every human being—implicitly a Jew—is to act rightly so as to enable their return to the source (Robinson 2021). The second path leads to the British-Irish actor and author Spike Milligan, who, being a dozen or so years older than Cohen, coined the following aphorism: 'Blessed are the cracked for they let in the light' (Adler 2021). Suffering from bipolar disorder himself, he consciously indicated a flaw as a mark of distinction rather than condemnation. The crack was to make it possible to see—principally in ourselves—that suffering can be forged into the inspiration to create. In this case, the light is not so much trapped as liberates the cracked. It is difficult to say which path Cohen was following—perhaps it was yet another—but I do believe that the two of them complement each other. This is because they pose a challenge consisting in the ceaseless working through oneself, so that despite adversities one may become better, nobler and thus save the world. It is also in both cases that the crack—a problem that needs to be repaired—is the leaven.

5.4 Cultural Shock

Above, I have presented cognitive results as a testimony to the individual's experience of a change. It was a rapid and irreversible change. Therefore, we may assume that the individual experienced a shock. Mandelstam, Nabokov and Cohen testified to the above from the point of view of writers—people referring to their individual experience and imagination. We could therefore place their experience at the crossroads of the common and the aesthetic perspectives. The issue of the shock is analysed simultaneously with reference to the scientific perspective or the relation between the scientific and the common. As in the case of the literary works under discussion, I am referring here to the level of individual experience. In communication science, this issue takes the form of a cultural shock and applies to the behavioural problem, which was previously regular, as it was referred to the domesticated surroundings. The term 'cultural shock' was coined in 1954 by the Canadian-American[2] economist and anthropologist Kalervo Oberg. It denoted an event 'precipitated by the anxiety that results from losing all our familiar signs and symbols of social intercourse' (Oberg 1960: 177). Anxiety was to appear in the situation in which the individual finds themselves in a culturally alien environment. Simultaneously, Oberg compared it to a health condition, the course of which may be thought to resemble the curvature of the letter U. It should be added that the course could be interrupted if the individual withdrew themselves at any stage from the alien environment.

The very figure of the U curve picturing the crisis of the individual experiencing a cultural shock is attributed to have been created in 1955 by the Norwegian sociologist Sverre Lysgaard (Kim and Ruben 1988: 302). Therefore, it seems that the two researchers simultaneously presented a similar scheme. Oberg's scheme underlined the emotional dimension of experience. This is important as the future researcher of communication, William Gudykunst, distinguished affective and cognitive processes (Gudykunst 1988). When writing about the possible effects of intercultural communication—and it should be added that he explored relations between groups rather than individuals—he pointed to the reduction of both anxiety and uncertainty. Hence,

---

[2] At the same time, he was of Finnish origin.

it should be remembered that emotional shock is one of the two sides of the same coin; that its other side—the cognitive one—should not be disregarded. After all, it applies to the figures of the act and the result as discussed above. Anyway, the U curve is to picture a health condition starting with the stage of the initial delight (honeymoon), through regression and gradual correction of behaviour (adjustment) to the final adaptation. On the one hand, we could presume that the final effect lies in the combating of the disease and a return to the state of balance. However, on the other hand, the final balance is tantamount to a cognitive rather than an emotional transformation. It seems that only some later researchers of the cultural shock such as Young Kim and Brent Ruben became aware of the above, since they decided that the variant assuming the U curve considers the cultural shock as a problem which needs to be solved rather than a catalyst of individual development (Kim and Ruben 1988).

Indeed, the original idea concerning the cultural shock assumed that the experience of an emotional disturbance may be overcome in order for the individual to adjust to the alien environment. It could therefore be suspected that Oberg perceived the state of balance resulting from the individual's stay in a familiar, well-known area as a starting state. What is baffling in this context is his understanding of culture, which was to include man-made physical objects, social institutions, ideas and beliefs. Therefore, it was about a conglomerate of specific, abstract and mental objects, with the last ones—beliefs—being counterparts of cognitive results rather than attitudes; Oberg understood attitudes as principally testimonies to emotional states. Examples include an aggressive attitude at the stage of regression or a superior attitude at the stage of the correction of behaviour. Culture understood in this way—in the context of understanding the individual in the target multicultural reality—allows to refer emotional disturbance to a relatively well-defined environment, allowing in consequence not only a diagnosis, but also treatment. This is why as much as the experience of a shock may be common to both a protagonist of the poet's (or writer's) story and an anthropologist, in the latter case the effects of this experience can be controlled in a potentially predictable way—even if this would consist in the adjustment to new conditions of life.

5.5 Stultitia

Cultural shock can be compared to another event, considered as a part of another field of knowledge. On the one hand, I am referring to the field of science, as a part of which the focus is not on the recording of individual emotional disturbances, but on the disturbances, which can be objectified, i.e. made a typical object of analysis. On the other hand, I wish to turn attention to the researcher's reference to themselves as a subject, which avoids adjustment to new, although already binding conditions. For this purpose, I am returning to the example of Faubion, Rabinow and Stavrianakis, who reflected on the triad of the present, the actual and the contemporary (Rabinow and Stavrianakis 2013). In particular, I am interested in the process of a breakdown of collaboration as a part of research projects in which the last two of them participated.[3] Let me recall that the problem lay in the divergence of the goals of the anthropologists and the engineers, as a result of which the former left the research teams. Rabinow and Stavrianakis described the move using two terms originating from the German language: *Ausgang* and *Haltung*. The former comes from the already quoted text by Kant and denotes an exit understood as a discontinuation of an uncritical following of intellectual guardians of all sorts. The latter was borrowed—through Frederic Jameson—from Bertolt Brecht, who gave a dramaturgic character to the common term 'attitude', to which I have already referred when writing about culture. What the German director had in mind was putting together an attitude as a mental object, and posture as a physical one. Let us imagine an actor, who, while taking on a physical shape, simultaneously (or thus) adopts a mental attitude.

Rabinow and Stavrianakis created a new term *Ausgangshaltung* so as to name a way of relating to the interrupted collaboration, turning attention to the simultaneously analytical and dramaturgic nature of their move. It was to take the form of an online Studio designed 'to give initial shape to experiences and protonarratives and render them capable of being objectified and further developed as anthropological topics' (Rabinow and Stavrianakis 2013: 43). The move was simultaneously a breakthrough; referring to the Greek term 'kairos', the American anthropologists turned attention to turning points, situations which enforce decisions of long-term irreversible effects. The question is what this breakthrough turning point was. It turns out that it was not the breakdown of collaboration, but something Rabinow and Stavrianakis called an affect, which was an example of the already quoted experience requiring objectification. This is what, by the way, was testified to by the abovementioned area of the actual. What was at play was therefore an event covering an emotional disturbance which was an effect of another event—noticing that the heretofore undertaken attempts at the construction of cognitive results are unsuccessful. This is because knowledge was the main motive behind the activity of the American anthropologists. Although it was a question of an accident that they were leaving the fieldwork—which was tantamount to the area of the said research project—in such unexpected circumstances, these circumstances made them devote 'a more precise attention to what kind of knowledge is produced through fieldwork and what happens to that knowledge once the fieldwork proper is over' (Rabinow and Stavrianakis 2013: 33).

Anyway, the event—an affect—which needed to be named to allow distancing and analysis—had a breakthrough character. In this place Rabinow and Stavrianakis referred to the Foucauldian term 'stultitia'. The original Latin word 'stultus' denotes a stupid person exposing themselves to danger and chaos, someone, who—to quote the French thinker—is tossed about by contingent representations of the world. Such an individual is '"open to the outside world in as much as he allows these representations to get mixed up in his own mind with his passions, his desires, ambition, mental habits,

---

[3] I wrote above that Rabinow participated in a biotechnological project, while Stavrianakis—in a nanotechnological one. Having withdrawn from them, they together testified to the consequences of their step.

illusions (...)'" (Rabinow and Stavrianakis 2013: 37). Such a person is changeable and limited, and their thought is muddled, as they are unable to distance themselves to the stream of experience. I feel like repeating that they are not in control of the relation between a part and the whole, although they are an extreme case of such a condition, as they act chaotically. As such, they turn out to represent emotional disturbance. This disturbance was called 'stultitia'. Rabinow and Stavrianakis found themselves trapped, and it turned out that they might leave the trap by mutually diagnosing the problem. Only this new collaboration generated the said web of the analytical and dramaturgic processes; the disturbance was objectified. Both of them found themselves in control of the relation between a part and the whole, which resulted in—as I have already pointed out—stepping from participant-observation to second-order participant-observation.

5.6 Falsification

So far, I have been presenting examples of changes of an emotional nature. Some of them were a report from an individual, including imaginary experience, while others took the form of typical events or events requiring objectification. I would now like to discuss the change marked by a cognitive nature, which also refers to the experience of the individual—this experience being quite clearly diagnosed. I am referring to a research practice encountered in the empirical sciences, as a part of which events of the physical world are explained with the help of laws or hypotheses. For the sake of simplification, I shall refer to these last theoretical sentences. They are an element of two forms of reasoning: deduction and induction, with the latter being a special case of reduction (Bochenski 1965). Making another simplification, I shall say that when talking about deductive reasoning, one may disregard sources used for the formulation of the hypothesis. In contrast, when inductive reasoning is at play, it is assumed that the hypothesis is formed by generalizing a property (or relation) attributed to the content of the observation sentence, with its reference being the event under explanation. At the same time, in both cases this observation sentence is a derivative of the hypothesis. Verification of the formed hypothesis, which adopts two forms: confirmation and falsification, is one of the key steps of the research practice in question. Verification lies in the checking of whether the hypothesis is consistent with the observation sentences recorded after it has been formed.

The difference between the two forms of verification is fundamental. The confirmation, which lies in proving the hypothesis, is potentially endless. This is because we may refer both to sentences which testify to current events, and to events from the past. On the other hand, falsification, i.e. a rejection of the hypothesis—let me simplify this representation again—is finite. The procedure of checking consists in the conjunctive collation of observation sentences. A sentence confirming the hypothesis is true, while a sentence which does not confirm it is false. A conjunction of even a single false sentence with any number of true sentences produces a false sentence. Hence, falsification is tantamount to a relatively finite and clear verdict.

What now emerges is a question of the consequences of this situation. The problem is that, first, the scope of the knowledge which adopts the form of a cognitive result is determined by the content of the hypothesis. Secondly, it is essentially the case that the researcher (and the community of other researchers and laymen after them) relies on a substantial set of hypotheses rather than a single hypothesis. However, in a given case, a given case may be focused on the checking of the usefulness of a single hypothesis rather than all of them. This case can be illustrated by the following formula: $[(H \wedge T) \rightarrow S] \wedge \sim S \rightarrow \sim H \vee \sim T$, with H symbolising the hypothesis, T—theory, i.e. the set of other hypotheses, and S—the observation sentence. This formula is a special case of *modus tollens*—a deductive rule, which—owing to the use of De Morgan's first law—pictures the possibility of rejecting a single hypothesis and keeping the rest of the cognitive results in the situation of the appearance of an empirical testimony inconsistent with the hypothesis freshly added to the set of the other hypotheses—the theory.

The presented formula allows the researcher to assume that the unwanted testimony does not necessarily have to ruin the heretofore constructed set of cognitive results, i.e. nullify the knowledge about the world collected so far (or its significant part). It thus, however, indicates that the knowledge is not ultimate. This seemingly banal remark has, however, its consequences for the moment in which the researcher assesses the result of the test and judges that it is adverse. This is because this moment witnesses, however small, a disturbance of the cognitive order. Until then, the researcher has relied on the content of the cognitive result, which from then on has to be abandoned. Let us not forget that the proposition takes place in the pragmatic domain. The cognitive order does not always depend on the syntactic and semantic characteristics of a given cognitive result. Among other things, the researcher acts: adopts an attitude to the hypothesis. They assess it, connect it with an intention, and hold beliefs. The cognitive order remains stable until a given attitude (or a bundle of them) changes its character. This change—understood as presented above when I was referring to Ajdukiewicz—makes the said order undergo a disturbance. The logical formula presented in the paragraph above is referred to in order to make the researcher aware that the disturbance does not have to be dramatic. Nevertheless, it can not only affect one's emotional condition, but also make one approach the formulation of general (theoretical) propositions on the world of contingent physical events with considerable caution.

5.7 In a Community

Above, I have presented various cases of the crack—an event which, firstly, testifies to the non-continuity of the world, and, secondly, is a manifestation of time. Both these elements are two sides of the same coin. Because of the dramatic nature of the crack, people may seek an escape from its further experience. Although this experience may in principle be emotional—as can be seen in particular in literary examples—they always affect the cognitive order. However, I have already underlined that knowledge can take two forms. Hence, I am interested not only in the order understood as a set of cognitive results—poems or hypotheses—but also a bundle of cognitive acts—recollections or utterances—which can be examples of a crack themselves. I simultaneously interpret these acts through a reference to attitudes, with which they are interrelated. Culture, which—let me recall—orders human actions, is a network of such attitudes. An imaginary model situation would lie in a socially shared network of attitudes, generated inductively owing to repetitive physical events, including human actions, mutually ordering the actions of members of a given group and other events of the physical world. The case I am referring to would lie in a disturbance of this mechanism by an event inconsistent with the anticipation adopted by the network of attitudes and an inadequate reaction to it.

Obviously, such a situation is a pure simplification, but it is to highlight the effect of a disruption of the continuity of the world and a testimony to an irreversible change. At the same time, from the point of view of an individual who experienced it, this situation—as indicated by the above examples—would require a repair or reconstruction. This is because it is impossible to persist in the state of a cognitive—but also an emotional—chaos. It is inter alia for this reason that Kołakowski turned attention to the significance of myth, and Geertz underlined the importance of a religious perspective. It was—in both cases—about the tool which may protect people—community members—against this chaos. Thus, the event as a crack or a disruption of the heretofore existing order of events turns out to result in the possibility to escape from the critical coping with a crisis. Myth and custom protect members of the group against investigating the possible causes of unpredictable events. This does not, however, mean that in the context of such a scenario people get rid of attitudes. They remain creatures of culture. However, owing to the tools mentioned above, these attitudes are configured in such a way as to neglect any testimony inconsistent with the anticipation offered by this tool. This is inter alia what the dogmatic nature of religious perspective is about. It is hence for this reason that myth is not a form of (scientific) explanation, although it explains a course of events. It thus becomes a worldview-related form—with the worldview understood as a normative rather than common-sensical knowledge.

The question is whether every form of worldview has a mythical nature. First, the worldview has an integrating character. It is therefore not an accident that myth is a useful tool which allows to successfully connect human groups. Secondly, integration may take different forms. For example, Durkheim distinguished between the collective and the individual conscience (Durkheim 1984). The former was typical for a traditional society, and the latter—for the modern one. It turns out, however, that the 'privatization' of the worldview was not tantamount to the breakdown of the group, but to its different—organic—configuration. What has remained the guiding principle was what the French sociologist called duty, which was not to lose its impact as a part of the modern society. Thirdly, when Kmita wrote about the relation between the symbolic and the technological sphere, he assumed that in the case of the industrial society, the symbolic sphere loses its social nature for the benefit of the technological sphere, which becomes the new worldview. Fourthly, we may assume following Rogers Brubaker that the group does not have a substantial nature; it is not so much a construct of the researcher as of pragmatically-oriented social engineers, who essentially generate people's sense of belongingness to a group through conflict (as a part of the 'self/other' relation) (Brubaker 2004). Summing up these elements, we may point out that where the group appears, a worldview is also present, which has a mythical nature in the sense that it determines standards of behaviour rather than allows to explain this behaviour. What is more, people are social animals inter alia because they very easily succumb to the impact of standards regardless of whether we are talking about members of the traditional or the modern society.[4]

5.8 On One's Own

Nevertheless, the worldview does not have the same impact on every individual. On the one hand, some individuals, at least temporarily, set themselves free from its influence, adopting the already discussed aesthetic perspective. This takes place in particular in the case of a cultural shock or stultitia. I have already written about this perspective without, out of necessity, referring to a borderline situation. Marlow's sensation of the pervasive power of the forest did not result from a traumatic experience. However, in the context of Mandelstam's or Cohen's representations, it is worth pointing out that the experience of suffering may lead to a suspension of someone's attitudes to the surrounding world. By the way, this is testified to by the said poets' works themselves. Should they be included in the worldview plan? Not necessarily and this is not because the plan is often exploited by the above-mentioned social engineers, who are thoroughly aware of the manipulatory potential of the pragmatic dimension of narration. This dimension makes another strategy possible. When someone reads about the smashed spine of a beast as a condition of the ending epoch, they do not have to—as these engineers would like to have it—joyfully welcome the approach of the new epoch, which promises heaven on earth. They thus do not have to succumb to the normative nature of the message. This individual may instead focus on images and sounds; construct, in their imagination, shapes and smells, at the same time experiencing sensations—for example, words reaching them in the rhythm intended by the poet.

Hence, during the said experience of stimuli, judging may be suspended for the benefit of the contemplation of the imaginary world and the world experienced with the senses. It is an interesting thing, although rather simple in the methodological sense, that the same result (or the same goal) may be reached through various causes (or various means). Conrad's sensation was positive, while Nabokov's one—negative. However, both of them could make the reader of their novels cease to reflect on the relation between a part and the whole, which, however, was not tantamount to the introduction of a normative element. The measure they introduced lay in the reader's ability to connect the imagined world with the real world and become a

---

[4] I also discussed the relation between the worldview and myth elsewhere, comparing, inter alia, the positions of Durkheim, Kołakowski and Kmita (Zaporowski 2020).

5.8 On One's Own

connector who juggles shapes and sounds, moving from one world to another. Such a state of unity could be, and usually is, temporary, but this temporariness was sufficient to release the reader from the often tragic sensations, in the context of which contemplation gave way to the making or the recalling of choices (propositions), including the awareness of the consequences of these acts. This measure may be attributed a therapeutic value. After all, reactions to a trauma may include taking attention away from the current course of events. What is also at play is the pedagogical value. Since suffering is not temporary and repeats itself, it is good to learn to react to it adequately, remembering to both maintain moderation and not run away from the world into fiction. Hence, reading calls for a withdrawal from the process of proposition, which is resumed in a while.

If, however, attitudes to events of the physical world would have to be adopted, then also individuals are at play—this time, they are forced to distance themselves to the group. Krug is an extreme example. The philosopher living in the world of dictatorship maintains his independence—although as it turns out, at the price of isolation and, ultimately, public execution. However, we may also present a researcher whose hypothesis becomes falsified. I am referring to an individual who challenges a group of other researchers, formulating a hypothesis of a possibly low probability,[5] and thus facing a specific risk of failure of their own project. If the hypothesis is falsified, they shall not obviously incur such an extreme trauma as the protagonist of Nabokov's novel, but they will have to make the effort of reformulating the system of their own attitudes contrary to other people. Importantly, as Rabinow put it, they will do it by themselves, but their work 'can never be done alone.' (Rabinow 2007: 49). Precisely because what is significant here is an independent work on self-transformation—rather than done alone—it is worth recalling Foucault's figure of the modern individual. This is because I do not think that a critical individual able to respond to the abovementioned breakdown can be found solely as a part of the scientific perspective associated mainly with the research achievements in the West between the 15th and the twentieth centuries.[6] Bearing in mind that according to the French philosopher the term 'modernity' should not be understood in terms of epochs, we are allowed to assume that in various spatiotemporal systems it is possible to tease out individuals who, having experienced a shock, including a cultural shock, can distance themselves to community worldviews, stepping beyond their cognitive boundaries.

---

[5] What I have in mind is the strategy proposed by Karl R. Popper, who criticised the project of probabilistic induction of representatives of the Vienna Circle, proposing a deductively-oriented project of falsificationism (Popper 1979). Cf. note 11 in Chapter 2.

[6] I am basically referring to the achievement commonly associated with individuals rather than groups.

5.9 A Couple

There is also the third possible reaction to a shock. What I have in mind is a pair of people who prompt each other to distance themselves to the consequences of the crisis of change. Examples include Rabinow and Stavrianakis who gave in to stultitia. Firstly, just like in the case of the writers, we are talking about an emotional disturbance. However, it does not lead to an escape from the problem and seeking refuge in the domain of myth. I am not referring to a narration created in an answer to the needs of members of a traditional or mass community. Let me say, recalling inter alia Kmita's position, that elite communities (such as those of scientists or engineers) being a part of the industrial (but also post-industrial) societies, also create their own worldviews. This lies in the symbolization of the technological sphere. Hence, this is not where Rabinow and Stavrianakis sought help against the consequences of the said affect. They decided to distance themselves to the impact of this worldview, dictating the goals, an example of which—amelioration of health and commercialization—I have presented above. The American anthropologists found help in mutual examination of the problem and mutual assistance in finding a solution to it. Secondly, such an act of social solidarity could not possibly not be tantamount to a reference to a worldview. I am not judging whether this move was indispensable. Nevertheless, it was owing to it that the two scholars could determine a joint goal for their further research. More importantly, this worldview was not encountered by or offered to them. It was co-created by them or initiated during a series of changes from the present to the contemporary.

Rabinow's and Stavrianakis's experience can be compared to Marcus's dialogical[7] manner of practising anthropology. The latter used the term 'collaboration' to testify to the mutual expansion of the field of knowledge with people from other professions and—more broadly—communities. This collaboration took the form of, inter alia, a joint book—e-mail correspondence—with a Portuguese marquis, Fernando Mascarenhas, the title of which grasped the essence of their joint experience (Marcus, Mascarenhas 2005). I am referring here to an opportunity for mutual development, which may but does not have to take place itself. This conditional nature of collaboration made it a series of both promising and risky events. Although Marcus and Mascarenhas did not experience—at least not immediately prior to or during their correspondence—such a penetrating shock as Rabinow and Stavrianakis, they yet showed that the path to opening oneself to new ways of relating to the world leads through the joint efforts of parties which did not yet share any specific experience. This situation reminds me of Donald Davidson's interpersonal communication project encountered in philosophy. In one of his works, he wrote about two people who begin to communicate for reasons other than already sharing the meaning of sentences or attitudes to utterances (Davidson 2001). The communication was tantamount to making one's utterance interpretable. The uttered sentence was yet to be understood, since the project assumed that the interlocutors had to 'calibrate' their

---

[7] I refer the term 'dialogical' to a type of ethnographic authority distinguished by James Clifford (Clifford 1983).

systems of attitudes (semantic structures) and in this way step beyond the heretofore existing scope of knowledge.

Mutual influence became the object of also my focus when I was discussing the problem of intercultural communication (Zaporowski 1999). I referred it to the multicultural world postulated in literature. Let me say, simplifying the picture, that culture was a kind of a monad such as I mentioned above, while the world in question was a series of such relatively stable, separate and coexisting constructs. At the same time, culture was often identified with the human community, due to which it was referred not only to race, as indicated for example by Lévi-Strauss, but also to ethnos, nation or organisation. I think that indications of the social nature of culture erased its key character: the ordering of the world of events, including one's own actions, referring to knowledge. Anyway, when posing a question of the possibility of communication between representatives of various cultures, I had to solve the problem of a mutual reference—the connecting—of constructs of different quality without the possibility of referring to an intermediating instance. In this place, I concluded that suggesting that members of the previously separate communities mutually learn the ordering systems of the other side would be an interesting way of solving the problem. In this way, they would establish a new community, the continued existence of which would depend on the two—implicitly: conjunctively— combined (ordering) systems. Thus, I wanted to show that one way to solve this problematic situation is to jointly create a previously non-existing system, which will allow both sides to share the thus-extended field of knowledge.

References

Adler K (2021) Spike Milligan: blessed are the cracked for they let in the light. http://www.academia.edu. Accessed 8 Nov 2021
Bochenski JM (1965) The methods of contemporary thought (trans Caws P). D. Reidel Pub. Co., Dordrecht
Brubaker R (2004) Ethnicity without groups. Harvard University Press, Cambridge, London
Clifford J (1983) On ethnographic authority. Representations 1(2):118–146
Cohen L (2021) Anthem (lyrics). http://www.azlyrics.com. Accessed 11 November 2021
Davidson D (2001) The second person. In: Davidson D (ed) Subjective, intersubjective, objective. Oxford University Press, Oxford, pp 107–121
Durkheim É (1984) [1893] The division of labour in society (trans Halls WD). Palgrave Macmillan, London
Gudykunst WB (1988) Uncertainty and anxiety. In: Kim YY, Gudykunst WB (eds) Theories in intercultural communication. SAGE Publications, Newbury Park, pp 123–156
Kim YY, Ruben BD (1988) Intercultural transformation. A systems theory. In: Kim YY, Gudykunst WB (eds) Theories in intercultural communication. SAGE Publications, Newbury Park, pp 299–321
Mandelstam O (2021) [1922] Century (trans Noon A). http://www.y112.wordpress.com. Accessed 11 Nov 2021
Marcus GE, Mascarenhas F (2005) Ocasião. The marquis and the anthropologist. A collaboration. AltaMira Press, Walnut Creek
Nabokov V (1974) [1947] Bend sinister. Penguin Books, Harmondsworth

Oberg C (1960) Cultural shock: adjustment to new cultural environments. Pract Anthropol 7:177–182

Popper KR (1979) Conjectural knowledge: my solution of the problem of induction. Popper KR objective knowledge. Oxford University Press, Oxford, pp 1–30

Proust M (2003) [1913] In search of lost time: the way by Swann's (trans Davis L). Penguin Books, London

Rabinow P, Stavrianakis A (2013) Demands of the day. On the logic of anthropological inquiry. The University of Chicago Press, Chicago, London

Rabinow P (2007) Marking time. On the anthropology of the contemporary. Princeton University Press, Princeton, Oxford

Rabinow P (2011) The accompaniment. Assembling the contemporary. The University of Chicago Press, Chicago, London

Robinson G (2021) Isaac Luria and kabbalah in Safed. http://www.myjewishlearning.com. Accessed 8 Nov 2021

Toker L (1989) Nabokov: the mystery of literary structures. Cornell University Press, Ithaca, London

Zaporowski A (1999) Is cross-cultural communication possible? In: Beitter UE (ed) The new europe at the Crossroads. Peter Lang Publishing, New York, pp 295–305

Zaporowski A (2014) Rozne oblicza wspolczesnosci (Various Faces of the Contemporary). Lud 98:299–307

Zaporowski A (2020) O pewnym przypadku relacyjnego ujęcia mitu (On a particular case of the relational approach to myth). In: Obrebska M, Pankalla A (eds) Mity kultury wspolczesnej. Perspektywa psychoantropologiczna (Myths of the contemporary culture. A psychological and anthropological perspective). Wydawnictwo Nauk Społecznych i Humanistycznych UAM, Poznan, pp 87–106

Chapter 6
Culture

Abstract It is here where I present my original concept of culture as a set of propositional attitudes to condition purposeful actions. Culture is thus attributed to individuals rather than communities. A human being is then a system that not only reacts to but also initiates changes in an ordered way. This is how culture fluctuates. Cultural difference is then discussed. I claim it is this difference with respect to which cultural fluctuation is possible but not necessary, and where the change is caused by human actions. Such actions provoke counterparts to alter their previous behavior. In this respect cultures may, but need not, be calibrated despite being changed. Since a crack is an irreversible event, it is confronted with a border, the crossing of which opens new perspectives.

Keywords Culture · Attitude · Action · Difference · Border

6.1 Different Worlds

When writing above about two people facing the challenge of intercultural communication, I am referring to the vocabulary which lies at the foundation not only of culture, but also intercultural relations. This vocabulary has been present in the humanities at least since the end of the nineteenth century; I think that it needs to be critically reformulated rather than rejected in view of its alleged anachronism. Firstly, the term 'culture' was defined in the categories of the integrating being at the same time the ordering. Among other things, Edward Tylor concluded in 1871 that culture is not only acquired in its social environment, but also constitutes a range of capabilities and habits (Tylor 1920), which stand behind the stability of the individual's actions. He thus suggested that these actions are integrated with the actions of other individuals in view of the community-based framework of ordering the world. This is manifested by such elements as knowledge, beliefs, law and customs. Tylor could not take into account Geertz's aesthetic perspective—after all, he lived in different times—so when he was referring to art, he understood it just like the abovementioned elements of culture, i.e. turned attention to its integrating nature. Secondly, the British anthropologist associated culture with a stable (implicitly: traditional) community. For this reason it could be concluded that these capabilities and habits are replicated from generation to generation. It is noteworthy that the understanding of culture as a web of the ordering and the integrating with the unchangeable led to its interpretation as the aforementioned monad persisting in time.

Thirdly, the term 'civilisation', suggesting a co-existence of groups which are developed to a smaller or greater extent, was forced out or pushed into the background. The introduction of the term 'culture' into the game constituted a multicultural world populated by a multitude of the said monads. Nevertheless, the said (intercultural) relation adopted the 'traditional—modern' form, with permanence and unchangeability being indicated by the former rather than the latter. However, over time, the principle of the permanence of a given order started to be challenged, as testified to, for instance, by the thesis on the mutation of every society, put forward by the already mentioned Lévi-Strauss (Lévi-Strauss 1952). The vocabulary assuming the existence of an unchangeable ordering and integrating order, which certainly encompasses the traditional and possibly the modern, therefore appears imprecise. It can be reformulated in such a way that it is possible to state that the culture is ordering and integrating, although valid, temporarily. Some elements of culture are more flexible than others, which also depends on the specificity of a given group. For example, let me state, in reference to my earlier comment, that in the West, since the fifteenth century, scientific knowledge—inter alia in the context of the scientific perspective—underwent the said mutations much faster than beliefs. Simultaneously, the art of the so-called modern European communities (mainly in the West) was marked by a gradual turn towards abstraction, which was not experienced by the art of the so-called traditional European communities (mainly in the East). Nevertheless, in each case a given element of culture was (or is) to unite the community and make the actions of its members mutually predictable.

It might seem that a reference to yet another element—that of custom—denies the above. After all, it allows to grasp the unchangeability of the order and integration in a possibly problem-free manner. Let us, however, refer to Ludwig Wittgenstein's concept of language games. Games are based on rules. Wittgenstein maintained that obedience to these rules has, among other things, a customary nature (Wittgenstein 1958: 81). It is this nature which guarantees mutual understanding of the moves of parties to an interaction; these moves are predictable. In this sense, every game is permanent and unchangeable. However, nothing prevents the formulation of a new rule and the creation of a new game. The said mutations can therefore be understood as replacing old games with new ones, which allows a flexible interpretation of the categories of order and integration. Hence, instead of recognising the custom inter alia as a testimony to the traditional, it is sufficient to treat each of the elements of Tylor's seemingly anachronistic definition of culture as an indicator of a temporarily binding order and social integration. However, even if the said vocabulary is reformulated, its user will face the problem of not only replacing the old order with a new one, but also the putting together of the orders which are contemporary to one another. This problem unavoidably presents itself always when the category of change comes into the game. At the moment, it takes the form of a situation as a part of which two people representing different cultures funding different worlds of physical events begin to interact with each other.

6.2 Not a Monad But a Human Being

Let me point out in this place that I am not going to use the term 'culture' as understood in Tylor's, even reformulated, vocabulary. Firstly, I agree inter alia with Rees, who rejected spatially understood intercultural relations fitting the scheme 'traditional—modern'. What is more, I consider temporally understood difference—be it between two emergent systems—as useless, since I consider them a new form of former cultures postulated by the diffusionists, as discussed in one of my previous works (Zaporowski 2006). In both cases—regardless of whether we are discussing a holistic or a hybrid system—we are talking about beings which are separate, and, being specific atoms, not necessarily testify to the relational understanding of the human being as a creature entering into many interactions with various—not only human—domains of the physical world. Secondly, culture understood in this way—as a set of knowledge, law, customs, etc.—can be basically analysed in the categories of cognitive results. In turn, as already pointed out, I personally support a reference of results to cognitive acts. This is inter alia about a critical attitude to approaching the former—for example scientific propositions or moral rules—as isolated constructs with an unchangeable content. A more processual approach to culture is more useful, as it allows for example the highlighting of its temporal dimension. Let me recall that time can be grasped owing to the form being an event encompassing change. Concrete or abstract cultural forms in the shape of solely separate and static cognitive results make it impossible to grasp this dimension.

What do I understand by culture, then? I have already written about attitudes as ways of assessing cognitive results. I also referred attitudes to cognitive acts, since the latter involve, inter alia, the adoption of attitudes. At the same time, I pointed out that mental constructs (and I understand 'mental' semantically rather than ontologically) are also at play. Without entangling myself in analysing reasons for which the term 'attitude' was associated with the term 'proposition',[1] I shall say that I am interested in the name of a special relation between the cognitive act and the cognitive result. Let me add that elements of this relation can be analysed individually, which is particularly important when we notice that the act has a functional nature, while the result—an objective one. An attitude can be adopted in relation to something. In this sense, the cognitive act is, inter alia, tantamount to adopting an attitude to the result. Without referring to the instance of the attitude, such a relation could not possibly be grasped. The act and the result would be treated separately. I believe that it is more precise to approach them relationally. Since a human being can be considered a bundle of actions—cognitive acts—who simultaneously produces tools—cognitive results—will it not be more appropriate to combine the two aspects? In this context, I

---

[1] We may assume that the term 'propositional attitude' was coined by Bertrand Russell.

am following my earlier position (Zaporowski 2006, 2018) that culture is a system of attitudes, including propositional ones, which determine people's non-reflex actions. I am thus assuming that culture combines acts with (their) results.

Taking into account the relevant research vocabulary, we may ask about the possibility of the identification of a reference to the term 'attitude' under discussion. Attitude can certainly be compared to a habit, as indicated by, for example, Tylor. I believe that in many cases human actions are habitual. However, this is not always so, and moreover this understanding of the word 'habitual' makes the human being an adaptational system rather than a system which initiates changes themselves. For this reason, inquiry—the second element of the relation, this time postulated by Dewey, would be an equally good candidate. However, in this case there is no reference to even be it a temporary permanence of a reaction to the problem encountered. Let us therefore assume that an attitude can be abstracted from a bundle of many actions undertaken by an individual during their many interactions with their surroundings, including other people. Wittgenstein wrote that words are tools (Wittgenstein 1958: 6). However, they would be useless if the individual—including a group member—could not use them. Since, then, attitudes are ways which allow manipulation of the said tools, may they reveal themselves through a testimony in the form of the effective action of the tools. It will then turn out that the individual is not an automaton reacting to a challenge on the basis of a ready algorithm, but a system of coordination of their own actions in an unpredictable environment with the help of a potentially fluctuating system of attitudes, which updates itself along with the changes of this environment.

6.3 Calibration of Cultures

In the context of the above words, intercultural difference is tantamount to a dissimilarity of at least two systems of attitudes. There are two aspects to this dissimilarity. First, a given system of attitudes is temporal. Since we are talking about beings which last for some time, a testimony to the change of this system of attitudes in the form of a single event is not an option. Such an event testifies to an adoption of an attitude, and thus is a testimony of an attitude being an element of a given configuration, which lasts. However, this can also be about the adoption of a new attitude or doing it in a different way,[2] which will be the case when an event will be compared to another—earlier—event. For example, if a pedestrian becomes a cyclist, they will be able to verify their attitude to the speed with which they cover a given distance, thus changing their belief (or a degree of belief) as to the previous practice. Moreover, the new practice may enrich the new belief (or a new degree of belief) with another, previously unknown attitude such as a desire to move even faster. Hence,

---

[2] The adoption of an attitude is gradable to various degrees. For example, a sentence can be considered true or false, but there are also many other possibilities between these two extremes. A sentence A can be considered more probable during time t than during time t^+.

only a comparison of at least two events, one of which, for example a physical one, consisting in the identification of the new situation with another, let us say mental event—a recalling of a former situation—will allow a statement that an attitude, its degree, or the state of configuration the attitude is an element of, has changed. At the same time, a change of configuration causes a change of the relation to the content with reference to which a given attitude is adopted, as discussed in my earlier work (Zaporowski 2009). For instance, if during time t attitude A creates a configuration with attitudes B and C, while during time t^+—a configuration with attitudes C and D, then the relation of A to the content during time t and t^+ is different. The said pedestrian, who until then had a positive attitude to walks, may—vide the experience of cycling—conclude that this is not the only option for a healthy lifestyle. Secondly, the said system of attitudes is referred to the individual, not a group. Of course, different people may share not only the particular attitudes, but also their configurations. This results from the potentially social nature of attitudes. However, I have already postulated that the individual may participate in many groups simultaneously (Zaporowski 2018). Learning to adopt the particular attitudes and doing it relationally in relation to other attitudes, the individual represents a certain starting cognitive potential being a function of the state of their organism, their experience of a specific environment, etc. The genetic structure, *qualia* and the surroundings make the process of the adoption of attitudes unique rather than typical. In this context, I consider a belief that the social nature of the learning of orientation in the physical world makes the individual a group animal too simplified. Rather, people try to maintain a balance between their individual leanings and group duties. On the one hand, society is an environment which cannot be disregarded easily, while on the other hand, individual experience does not allow this environment to rule uncritically. The above may be testified to by Bronisław Malinowski's reflections on Melanesians' practical rather than dogmatic attitude to themselves and the group (by the way, they were similar in this respect to their contemporary Europeans; Malinowski 1985). Hence, I assume that culture as a system of attitudes is a socially available ordering system, which can be individually modelled through interaction with not only others, but also the world of events as such.

We may therefore imagine a meeting of two individuals who do not share their systems of attitudes. What is more, the difference between the individuals' systems is gradable, and the meeting itself should be approached processually. May one case lie in the total and mutual incomprehensibility of the partners' actions. Can a Melanesian understand what a Sicilian may use snow for? May another case lie in one of the parties understanding the other one better than the other way round. Let us imagine a layman, who, without knowing much more about water than their senses dictate them, meets an expert familiar with its chemical structure. Another variant can involve a small difference in perspectives, when a person who swam in the Mediterranean Sea meets a person swimming in the Baltic Sea and wants to check the taste of seawater. I understand differences in the systems of attitudes determining the particular actions as a starting point for the process of the mutual transformation of the configuration of the two systems. It is similar in the case of people's contact with events of the physical world as such, although in this case this is not necessarily about mutual comprehensibility,[3] but the individual's ability to appropriately react to an external stimulus. Just like people learn, inter alia, to adopt attitudes socially, they are also able to change these attitudes—and their systems—in the face of testimonies in the form of physical events, including the actions of partners to interactions. Bringing two previously different systems closer to each other to such an extent that the partner's actions are considered one's own can be one of the effects of the process of the mutual identification of action, i.e. the determination of the relation between action and the system of attitudes behind it. I refer to this effect as a calibration of cultures.[4]

6.4 What if There is No Calibration?

The said calibration of cultures is one of the effects of interactions between people. On the one hand, it is rare, as it assumes the acknowledgement of the partner's action as a sufficient and necessary condition of one's own action. Such a balance is subject to the impact of external conditions. This is because we must not forget that the individual interacts with many partners and—generally speaking—events of the physical world at the same time, while as a part of the relational understanding of their actions, a change in one area of contact potentially invokes a change in every other area. On the other hand, calibration may happen periodically, and subsequently give way to the separation of cultures, which can be temporary and gradable. Such fluctuation is characteristic for interactions of the possibly flexible and creative individuals, and can be contrasted with the situation of a relatively stable relation based on a repetition of the same scheme. Examples may include actions referring to customs or beliefs. If, then, we are talking about intercultural contact, the range of the possible results of failure to communicate between individuals spreads between two extremes. The first one lies in the possibly constant change of configuration of one of the parties (or both of them) in reaction to the partner's actions, with a simultaneous maintenance of a relevant degree of otherness. The second extreme is tantamount to the maintenance of otherness by the possible resistance of one of the parties (or both of them) to a change of configuration of attitudes in the light of the partner's actions.

A change or resistance against a change of configuration of attitudes puts the issue of the above-mentioned cognitive results in a special light. Since culture determines a bundle of individual actions, they may be called cultural actions. They will therefore differ from, for example, actions conditioned by the reflex arc. Cultural actions will simultaneously be cognitive acts involving the adoption of an attitude. I shall add that since attitudes create a configuration, not just a single attitude, all the attitudes making up the configuration will be adopted. What is a cognitive result then? It turns

---

[3] Strictly speaking, the gradation of mutual understanding also comes into play here. The impact of inanimate things on people is different than the impact of living things.

[4] I believe that this is consistent with the scenario of Davidson, who analysed interpersonal relations in the context of the relation between prior theory and passing theory (Davidson 2005).

out to be a cultural work, since it is constructed as an effect of cognitive acts—cultural actions—to then become a point of reference for the subsequent actions, according to, for example, the above-described procedure of inductive or deductive reasoning. A cultural work will be useful for an individual—it will for instance constitute a cognitive tool—the afore-mentioned words by Wittgenstein—as much as the individual adopts an attitude (a configuration of attitudes) determining the actions owing to which the work has been constructed. For example, we should be positively convinced that since all the ravens we see are black, every raven encountered is black, while at the same time assessing that it is possible to see a raven of another colour. After all, inductive reasoning is not unfailingly logical.

Therefore, the formula '$(A \rightarrow B) \wedge B \rightarrow A$' shall not be useful for someone who, inter alia, does not conclude that if A is false, while B—true, the entire formula is false. This particular individual shall recognise that this is so, if they refer to an instance (in this case—a belief), resulting from practice, that one of the possibilities is appropriate in a given case. The act of recognition therefore suggests that culture is different than cultural work, while simultaneously that works cannot be identified without reference to culture. This is because on the one hand, the work would not have been created without the ability to adopt attitudes, while on the other—an absence of this ability makes the work useless or used wrongly. In this context, the individual's encounter with changing or previously unknown surroundings enforces their ceaseless updating of their configuration of attitudes, which allows them not to become lost in these surroundings. However, this applies to the ones who are flexible and creative enough to find the courage to give up the heretofore existing culture. The permanence of many ways in which worldview-related narrations are used testifies to the fact that many people resist the challenge of the changing surroundings; this resistance is a group rather than an individual phenomenon. Hence, it is not accidental that when analysing the development of intercultural competences, Kim and Ruben attributed them to individuals rather than communities (Kim and Ruben 1988). In this sense, as much as the intercultural contact does not have to lead to intercultural communication, it is basically—at least in relation to individuals—a catalyst of changes of one's own culture.

6.5 Foucault's Problem

Writing about the cultural change may, however, appear problematic. After all, so far I have understood change as a relation taking place within the whole—an event. The event covers state A followed by state non-A. In turn, I pointed out that I understand culture as a lasting configuration of attitudes. Hence, how to present the relation of the succession of configuration Y after configuration X, when I assume that each of them is a whole? Let me recall that I have already discussed the event as a testimony to an attitude (and, further, configuration), including a new attitude. Firstly, it comes into the game when, for example, the interpreter recognises that the acting individual adopted an attitude, i.e. commenced a cultural rather than a reflex action. The interpreter does not yet have to know whether this is about a new configuration of attitudes or not. Anyway, let me repeat that individual action does not force the interpreter to adopt their own attitude—the said recognition—towards any of the configurations. Only a bundle of actions referring to the situation identifiable by the interpreter may mean that one of the attitudes is materializing itself. Another one will follow. It will be necessary to assess which bundles of actions appearing increasingly less regularly are giving way to other bundles appearing increasingly more regularly, rather than coexist with them, which could suggest that both bundles complement one another. Hence, the term 'change' is a name for a relation between the receding regularity of some actions with the simultaneous increase in the regularity of other actions, each of which is determined by different systems of attitudes.

In this context, let us have a look at Foucault's modern human being. Firstly, what the French philosopher had in mind was the individual stepping beyond the social practice of potentially uncritical people. Secondly, the individual experience was understood processually. The individual adopts the attitude of modernity, which is potentially replicated in new conditions. The ethos Foucault referred to was tantamount to ceaseless work on oneself, unending self-transformation in the critical mode, giving a new form to what is referred to as human nature in the possibly individual and distinct categories. What is interesting is the initiating movement—as a bundle of actions—since it is tantamount to the adoption of a new attitude; by differentiating between the old and the new practice, the interpreter may see change consisting in the individual's differentiation from their peers. This step becomes routinised as much as the necessity of adopting a critical attitude to oneself emerges again in the changed conditions. Therefore, Foucault's modern individual is someone who testifies to their ethos through the process of the adoption of an increasingly new course of critique of, inter alia, their own self. However, the very criticism, as much as a given individual intends to remain modern, does not change, since its only alternative is an absence of criticism. It is a consequence of putting a modern individual together with an anti-modern one.

However, cannot one be less or more critical today than they were yesterday? Is the change not perceivable in this case? It is actually so, and it should be remembered that this process assumes oscillation between the extreme cases of a single sequence.[5] The attitude of modernity is tailored in such a way that it can be contrasted with the absence of an attitude rather than an alternative attitude. What is more, Foucault indicated a single attitude and hence the relation between the modern and the anti-modern individual turns out to be rather clear and simple. It is different in the case of a relation between two people who do not share a culture. I am referring to the situation when these individuals are modern, but when simultaneously their culture is composed of some other, mutually incommensurable attitudes. The question is

---

[5] In this context, a special place among attitudes is adopted by a belief which is not determined on the scale from the lowest to the highest degree. For example, considering a sentence true or false (or less or more probable) indicates that extremes testify to the adoption of an attitude equally clearly. In this case, change is always fundamental. This is what I am discussing below.

whether the attitude to oneself as established by the Foucauldian ethos changes when it is accompanied by a configuration of other attitudes. If the 'self' is the name of an object to which modernity refers in the same way as propositional attitudes refer to a proposition, then, in consistence with what I have written above, adding other attitudes to the said ethos changes this object's attitude to itself; it is about a different 'self' than earlier. In this sense, two modern individuals may mutually not recognise themselves as such; they may remain alien to each other. Therefore, we may imagine a situation when criticality to oneself not only distances someone from the community of anti-modern people, but also does not make them closer to the people who are as modern as they are.

6.6 Kmita's Problem

The example of Rabinow and Stavrianakis, who supported each other to withdraw themselves from their projects—field studies—shows the role played not only by the modern attitude, i.e. being critical to the current practice, but also a collection of propositions ordering that this practice be discontinued. What do I mean? On the one hand, the said ethos, which, generally speaking, makes the individual find distance to themselves, is an element of culture. On the other hand, the said propositions are cultural works. However, the individual's relation to these works is also significant. And it is not the attitude of modernity which is responsible for this relation. The latter is used for the criticism of oneself. The said propositions can be acknowledged or rejected. Each of the alternatives will become a component of cultural action determined by a particular belief.[6] In view of the above, in the example in question, culture will be simultaneously composed of modernity and a belief determining action (acknowledging or rejecting a proposition); the nature of the relation between these attitudes—such as a result or a conjunction—being secondary. Anyway, it can be assumed that at least temporarily Rabinow and Stavrianakis shared the same culture, simultaneously using the same tools—cultural works. Obviously, I am now referring to normative propositions, but they derive from propositions as such; the latter testify to the semantic domain. When writing these words, I am following Wittgenstein's indication that the semantic domain is subordinated to the pragmatic domain.[7] It is for this reason that a proposition becomes a tool as much as one knows how to use it. And this is the object of culture.

However, it turns out that the issue of cultural works is more complex. As discussed above, what Kmita had in mind when writing about culture was, inter alia, propositions. This is because when differentiating between normative and directive-like

---

[6] The difference between acknowledgement and rejection lies in the fact that there is another culture at play. This remark does not mean that I have adopted a relativist position. I am discussing this problem below.

[7] I am referring to the reflection that the meaning of words most often depends on their use (Wittgenstein 1958: 20).

propositions, he attempted to establish a relation between the goal of action and action as a means to reach the goal. At the same time, he pointed out that propositions are respected and acknowledged by members of a given group. He therefore pointed to the semantic and pragmatic domains, but he considered them together. There was no question of a proposition being rejected; a proposition had to be acknowledged by necessity. Hence, there was also no question of the condition of the acknowledgement of the proposition in the form of the instance of the attitude which can be adopted in a variety of ways. Anyway, let me recall that Kmita assumed a split of two spheres of culture—the symbolic and the technological. The former was used for integration, while the latter—for a manipulation of nature. Hence, it turned out that during social transformations—at the stage of religious communities which followed magical communities—two types of propositions were distinguished, with only the normative ones being acknowledged as a part of the technological sphere, and both the normative and the directive-like ones acknowledged as a part of the symbolic sphere. This is because Kmita suggested that the manipulation of nature has a reproductive character, and hence the individual does not necessarily have to use symbolic tools. And among other things this is what differs watering plants from festive celebrations.

However, when the split of the spheres in question has already happened, can the technological sphere at all last independently or can the domination of one symbolic sphere give way to the domination of another sphere? I am referring to a situation when the heretofore existing way of social integration recedes. Although Kmita assumed that in practice the technological sphere is determined by the symbolic sphere, and that they hence make a whole, he did not deny that the symbolic sphere can vanish. In such a situation, the normative proposition of the technological sphere would have to be acknowledged regardless of the acknowledgement of any normative proposition of the former symbolic sphere. According to Kmita, in such case a society would appear whose members would mythologise technology, as the previous sphere regulating technology with the help of myths-worldviews would succumb to atrophy, and the sphere which was until then subordinate to it would have to take over its role—if of course social integration was to be maintained. Let us imagine, for example, that the action of watering plants is not designed for aesthetic purposes, but is a goal in itself.[8] However, the second possibility appears, which Kmita failed to mention: the technological sphere may become an object of dispute between two different symbolic spheres. For instance, Mike McCormack in his novel *Solar Bones* presents a scene in which the protagonist, Marcus Conway, a contemporary engineer (who, by the way, was previously a student of a seminary), sees an exhibition of tools of torture (McCormack 2017). Hence, let us ask whether the engineer's use of a hammer aimed at the removal of an obstacle hampering the life of the local community[9] is identical with the use of the hammer by the executioner breaking a victim of the

---

[8] As a part of the functional vocabulary, this is tantamount to the replacement of the goal with its means.

[9] The said engineer deals with the repair of roads and bridges for the local county.

6.7 The Border as a Landmark

Therefore, it turns out that it is not only the symbolic sphere (the worldview, myth) which controls the technological sphere, i.e. the way of intervening into the world of physical events. We should also remember that a single technological sphere can be the subject of intervention of various symbolic spheres. What is more, we should not forget that what is important is not only the cultural works in the form of relevant propositions, but a system of attitudes—culture—which allows people to use these works. I therefore do not want to say that I am interested in the acknowledgement of propositions as such—I am interested in the relation between the acknowledgement (or rejection) and its object. In other words, I am posing a question what cultural action comes into play when the individual faces a specific cultural work, which can be their tool, if they at all recognise something as a tool. What I find significant in Kmita's context is the differentiation between two domains of cultural works, i.e. the spheres under discussion: the symbolic one and the technological one. What can be called a breakthrough in Foucault's context is the adoption of the attitude of modernity. In both cases we are dealing with an indelible crack, which is as much a testimony to time as a guideline that the human self—at group level from Kmita's point of view and at the individual level in the case of Foucault—adopts a new form. With reference to the last question, let me point out that the said crack appears when a heretofore accepted proposition is rejected. I am referring here to a situation when after a bundle of cultural actions which confirm a given cultural work, a new bundle emerges, negating this particular work.[10]

The question is whether the confirmed and the negated work is the same object. If it was so, I think that we would fall into the trap of cultural relativism as inter alia discussed by Geertz. According to the American anthropologist, this position lies in the acknowledgement of the existence of a universal value, with reference to which extreme positions can be taken (Geertz 1973a). A relativist is a person postulating independence of the object of reference from the manner of its description. I believe that the work in question may be distinguished only on a working basis, for the purposes of analyses, as a part of research vocabulary. As a tool of an acting individual, the proposition cannot be separated from the act of acknowledgement or rejection. In view of the above, for the interpreter, the object in the form of the acknowledged cultural work is not the same as the object in the form of negated cultural work. In this context, the crack I am discussing above differentiates between two worlds: the old one and the new one. At the same time, the border also applies to the worlds, which are contemporary (in the single, ahistorical sense indicated by

---

[10] A reverse situation is also possible. Basically then, I am referring to a discontinuation of the heretofore existing practice.

Rabinow). The word 'border' is not used here accidentally. The term 'crack', the meaning of which is similar, basically testifies to the temporal dimension; it has an eventful character. Border has its own permanence and it can also be understood spatially. For this reason, in particular contemporary cultures in the above-discussed sense border with each other, although there has not necessarily been a crack between them. Nevertheless, as suggested by, inter alia, Foucault, the border may be a signal of change.

This is why let us assume, remembering not only about the context of the contemporaneity of cultures, but also their change, that the border is perceived spatiotemporally. In this sense, it emerges or becomes so as to, among other things, divide something, which until that moment was a whole. It then can be connected with a testimony in the form of the aforementioned smashed spine or broken vessel. In *Solar Bones*, as mentioned above, Conway learns that his wife most probably—as can be assumed from a pregnancy test—is expecting a child. In his monologue, he says '(…) that's what the clear line through its little window was saying, definite as any line drawn in the sand or any surveyor's contour or any of those global parallels longitude and latitude which demark those national borders that are drawn up (…)' (McCormack 2017: 37). Conway then points out that this line was a sign of a threshold, which, once stepped beyond, stayed behind the young married couple forever, and that they now entered, perhaps unsure and unprepared, into a new world their child will be a part of. This line, or, better perhaps, its appearance, allows to notice the contrast, which makes orientation possible. In the *Heart of Darkness,* Marlow recalls that as a child he enjoyed looking at the map of Africa, which gradually, along with the subsequent discoveries, began to be filled with marks. In this case, the experience of becoming familiar with the surroundings, which step by step start to be covered with lines and dots—orientation marks being points of reference in something which until then was impenetrable—comes to the fore even more strongly.

6.8 Changeability of the Border

A line or a series of dots are signs of the said borders on maps, while barriers, posts, fences or walls—signs in fields, towns, and cities, along rivers and mountain ridges. Therefore, they are basically understood spatially. They are placed purposefully so as to—just like in the case of mythical tales—blur time. Their creators would like, in the context of different, often contrasting intentions, to guard their necessary and uninterrupted persistence. May the Berlin Wall be used as an example here. At the beginning of the twenty-first century, it is now something obsolete, but for almost thirty years of the second half of the twentieth century, it was a sinister sign not only of division, but also death; a sign, which at the time seemed to have no prospects for being blurred in the foreseeable future. On the other hand, the boundary marker indicates a place where the homeland ends and an alien land begins, and where separation, although not as disturbing as before, is equally indelible. This is by no means only about state borders, but about the separation of areas as such. There is

6.8 Changeability of the Border

a certain amusing, but also pitiful custom related to seaside sunbathing, which is popular among Poles. It consists in putting up beach windbreaks. This is not solely or primarily about protection against the wind, but separating oneself from other beach-goers. In turn, as much as a post or a fence is a symbol placed in the not necessarily symbolic space, a line on the map is a part of a symbolic system. In this sense, it represents a more advanced way of ordering the surroundings, just like inscriptions made on clay tablets or on the screen of tablet computers rather than on a stone lying nearby.

However, this more advanced form makes it possible to look at the boundary marker in a new way. When I wrote about the boundary in the context of cultural identity (Zaporowski 2009), I referred to the example of Poland, the shape of which changed in the past. Inter alia, I stated that on the one hand, the boundary divides two parts, but on the other, connects them. I also recall that I observed the changing of areas on many maps, which showed lines and dots running increasingly differently. Wandering along a river or a mountain ridge does not allow us to notice the temporariness of the encountered markers. It is only possible through a comparison of the present with the past. For instance, in the Island of Usedom by the Baltic Sea, you can walk or cycle between the Polish Swinoujscie and the German Ahlbeck, down a comfortable path running almost directly by the sea. At a certain stretch of the path, there is a several-dozen meters wide stretch of cleared forest, although one can see that no one cultivates land here. Perhaps in several dozen years, the forest will regrow. However, several decades ago, there was a border fence on both sides of the stretch, which at the time was ploughed. Only by recalling our own picture of those years or information from some handbook or encyclopaedia can we understand the change which has taken place here. The border remained, but its nature is now different than it was: it is open and connects new states. The map is a tool allowing us to follow changes much more effectively. The tool makes it possible for us to see that borders are never permanent. Similarly, the ground we tread on while walking along a mountain ridge or down the beach, is not permanent, since deep underground, tectonic plates are constantly moving.

A statement that the border is changeable is not tantamount to acknowledging that it can be erased. Let me repeat that is should be perceived in spatiotemporal terms. The borders emerge and vanish; they run in one place only to be delineated in another. Although the division of territories is basically temporal, it is a necessary condition for their identification, just like in the case of a relation between the truth and falsehood, the left and the right, the up and the down, etc. What is significant in the case of the inevitability of the border as a factor identifying the separated or the delimited object, which is of my interest here, is its mobility. It allows to approach the object in question as something flexible and transforming, something—to quote Cassirer—undergoing metamorphoses. At the same time, the irremovability of the border does not boil down to a post, fence, or line on the map. These objects are cultural works the use of which can be shared by individuals dispersed in a given space–time and in this sense they resemble tracks left by territorial animals. After all, we should not forget that a border is—just like change—a relation. Although, to repeat myself, there is a tendency to understand it in spatial categories, it is even then also a tool of thinking allowing an identification of things, which are no longer the same things in another spatiotemporal system. In this sense, the use of the term 'no longer' is not accidental.

6.9 Performative Nature of the Border

I wrote above that the border and the change are relations—but they are different ones. As much as the change refers to a succession and in this sense is a measure of time, the border testifies to an opposite, where time can be ignored. However, the opposite does not have to assume coexistence. In view of the above, let us assume that the border is associated with an end. I am talking about a processually understood reference to a point beyond which something unknown is lurking, or perhaps there is nothing at all. Let us focus here on the end of life—death. If, however, death is not tantamount to the end, as in the song *Death Is Not the End* (Dylan 2021), a new world, a new game or a new stage begins beyond it. In this context, the border already assumes the dimension of time and in this way we may say that it assumes change as a sufficient condition. There are two possibilities here. First, the border represented by the said lines on the map or posts by the river runs in increasingly different places, dividing territories or estates in an ever new way. Secondly, the border is expressed not only by the result of a pregnancy test, but also by the absence of effectiveness of the practice existing until this moment divides the old and the new possibilities for action. This last case is interesting in the context of the task which Foucault gave to the contemporary individual. I am referring to the crossing of the border, which is not something that happens, something along the lines of a divine intervention or a stroke of fate. I am talking about movement, which is conditioned by a specific attitude.

The term 'crossing' is not necessarily tantamount to the neighbouring of my culture with another culture. When Foucault, while writing about the contemporary attitude, mentioned that it had a boundary character, he suggested that the individual has to step beyond themselves rather than meet the other, waiting beyond this border. However, experience teaches that such an individual does not act single-handedly and often deals with people who order the world in a different way. In this sense, it can be assumed that stepping beyond one culture is tantamount to encountering a different culture. This does not necessarily take place on the border between some territories, but between bundles of actions determined by attitudes. In this sense, the border between cultures is abstract rather than specific. However, the field of actions looks different. On the one hand, they are not cognitive results available solely in space. On the other, cognitive acts are not only elements of the relation of succession, which testifies to the temporal dimension. In contrast to mental actions, physical events constitute an empirical testimony grasped in the space–time. Hence, cultural actions may testify to the intercultural border in the categories of the specific. The above will happen if we acknowledge that one system of attitudes is crossed in reaction to actions determined by another system of attitudes; that the order of one system is disturbed by external events—cultural actions.

Not all the borders are crossed. In this context, I may indicate a number of examples of individuals resisting the impact of the actions of others. Such individuals are usually associated with traditional societies. Hence, a statement that members of contemporary societies are more open to changes in relation to their culture does not come as a surprise. I do believe, however, that the opposition 'traditional/contemporary' is too simplified to describe the cultural character of the particular individuals. This is because this character does not depend solely on the social factor. As far as the latter is concerned, let me recall that according to Lévi-Strauss's interpretation, there is no opposition between the static and the cumulative history; the relation is gradable. In turn, Rabinow assumed that tradition and modernity are attitudes rather than social forms, which additionally—constituting the contemporaneity—have a dynamic nature. Hence, I believe that it is more appropriate to connect the cultural border with the acting individuals rather than groups. In connection with the above, I assume that on the one end of the possible situations we would deal with a situation involving at least a single individual not making changes to their culture. In turn, the possibility of a ceaseless transformation of one's system of attitudes by any party would come into play on the other end. Hence, a change—perhaps mutual—of one's own culture, indicating a fluctuation of the intercultural border, would take place between these extremes. The border of one's own culture perceived so far would give way to another border, acquiring a performative character, since it would be understood in the context of the interaction of individuals maintaining different grades of otherness. Action would testify to the dynamic nature of the border, while otherness—to the challenge of noticing one's own limitations.

References

Davidson D (2005) A nice derangement of epitaphs. In: Davidson D (ed) Language, truth, and history. Oxford University Press, Oxford, pp 89–108
Dylan B (2021) Death is not the end" (lyrics). http://www.genius.com. Accessed 11 Nov 2021
Geertz C (1973) The impact of the concept of culture on the concept of man. In: Geertz C (ed) The interpretation of cultures. Basic Books, New York, pp 33–54
Kim YY, Ruben BD (1988) Intercultural transformation. A systems theory. In: Kim YY, Gudykunst WB (eds) Theories in intercultural communication. SAGE Publications, Newbury Park, pp 299–321
Lévi-Strauss C (1952) Race and history. UNESCO, Paris
Malinowski B (1985) [1926] Crime and custom in savage society, Rowman & Allanheld, Totowa
McCormack M (2017) Solar bones. Canongate, Edinburgh
Tylor EB (1920) [1871] Primitive culture. Researches into the development of mythology, philosophy, religion, language, art, and custom 1. In: John Murray GP (ed) Putnam's Sons, London, New York
Wittgenstein L (1958) [1953] Philosophical investigations (trans Anscombe GEM). Basil Blackwell, Oxford
Zaporowski A (2009) Border and cultural identity. In: Ropo E, Tero A (eds) International conversations on curriculum studies. Sense Publishers, Rotterdam, Boston, Taipei, pp 327–337

Zaporowski A (2018) Action, belief, and community (trans Moroz-Darska A). Peter Lang GmbH, Berlin

Zaporowski A (2006) Czy komunikacja miedzykulturowa jest mozliwa? Strategia kulturoznawcza (Is Cross-Cultural Communication Possible? A Cultural and Critical Strategy). Wydawnictwo Naukowe UAM, Poznan

Chapter 7
Proposal

Abstract The final chapter is devoted to my proposal regarding managing the crack. The basic idea is to think of a movement in terms of a cycle. Unlike linear movement a crack represents time more accurately. On the other hand, the cyclical verification of data need not be associated with the repetition implied by using tools such as custom. This cycle may be disturbed and it is the crack which demonstrates this. Once one notices this one may cross the border to use another cyclical tool to order the world in a new and more effective way. So I propose viewing the world cyclically in time and self-critically being able to verify its usefulness. It is one's culture that allows one to reorient where the crack is overcome by one's reconfiguration of the set of attitudes.

Keywords Crack · Cycle · Time · Tool · Culture

7.1 The Border and Culture

The crossing of the border is tantamount to disturbing the heretofore existing order. The first signal is the experience of a crack. As I have already pointed out, it may be surprising, even dramatic. This is why the worldview blurring the consequences of such experience is so important. This can be exemplified by the acquisition of consciousness by people—as a species rather than individuals—as testified to by the myth on the expulsion of Adam and Eve from Eden. On the other hand, a multicultural person, as discussed by Kim and Ruben, intentionally deals with cultural shock, so as to navigate between mass or traditional communities with an even greater ease. In this case, the severity of the shock is the less intense the more often such a shock is experienced. At the same time, another circumstance springs up. It can be assumed that the emergence of consciousness—this time also at the individual level—changes one's attitude to the past in a different way than the crack accompanied by the awareness of one's own actions. In this last case, crossing the border is tantamount to perceiving the previous ordering of the world as a step to the current one. However, I do believe that it is impossible to determine some point zero, which marks the appearance of a subject gifted with consciousness; such a point is suggested by myth so as to establish the human existential condition in a possibly vivid manner—using images rather than concepts. I myself understand the crossing of subsequent borders processually, considering the human being a system, which develops, following others, to then—perhaps—construct their own cognitive tools.

In this sense, I distance myself to Hall's differentiation between type A and type B culture. Let me recall that the American theoretician of communication linked the former with learning, and the latter with acquiring. He hence separated the domain of consciousness from the domain of unconsciousness. According to Tylor, elements of culture are acquired, which reflects the common intuition that enculturation begins in infancy. However, people initially acquire knowledge or custom so as to subsequently learn how to use them. In this context, we should assume that culture A is a development of culture B rather than its opposite. In this sense, culture—which I understand as a potentially fluctuating system of attitudes—enables us to control the said crack. We should look at this relationally, i.e. refer both to cultural actions and cultural works. On the one hand, an event assumes a change, with a crack being one of the types of the latter. Hence, it is worth looking at the cultural actions testifying to the above. I am referring here in particular to the actions which testify to something new, not yet experienced. As already pointed out, a crack can be surprising, and hence tools alleviating consequences of such a surprise are constructed. It is however also possible that the very crack performs this role. I am referring to the metaphor. Following Davidson (Davidson 1978), I understand it performatively; I pay attention to its use regulating the meaning of the words it contains. This is because we need to differentiate between the proposition, which testifies to a collision of the possibly incommensurable semantic domains, and the act such as uttering this proposition with a view to making an intended impact on the interlocutor.[1]

On the other hand, metaphorisation often leads to the emergence of new cultural works, with myths being perhaps their most vivid examples. However, as a part of other perspectives, including the scientific one, we also encounter effects of the use of metaphors. If we look at the vocabulary of the science of electricity on the one hand, and that of sociology on the other, we may see that the formulas of both these branches are rooted in the vocabulary of, respectively, hydraulics and biology. The devil is a creative transformation of the male goat in the same sense, as the electric current—of the current of a watercourse (such as a river), while the society—of an organism. Culture enables a control of the crack or generates it, as it is a system reacting to the real or possible disturbances of the field—also the cognitive one. This is why the culturally acting individual is able to creatively use old works to construct new ones. Hence, stepping beyond the border is tantamount to the conversion of

---

[1] A proposition 'Your bag is as heavy as a stone' not only testifies to putting together incommensurable semantic domains, but can also be *uttered* in a serious or sneering tone. Similarly, a cognitively worthless proposition 'You have been possessed by the devil' *used* by an inquisitor bearing judicial powers has a different weight than when uttered by one's wife, who is helpless in view of the behaviour of her husband, watching a football match on TV. Simultaneously, there is no question here of the so-called metaphorical meaning. Elements of propositions have meanings determined by the belongingness of their references to relevant domains.

the old form of this world into a new one rather than to the breaking with the old world, to rejecting the Wittgensteinian ladder (Wittgenstein 2001). On the other hand, interacting individuals may represent cultures assuming a different use—for example, the old and the new—of the same works. This is, by the way, one of the reasons behind social tension. Anyway, without a system of attitudes, the individual would be a creature acting solely reactively. Culture allows them to act in advance.

7.2 The Metaphor and Verification

The use of metaphor is not the only way of constructing cultural works, although it indicates people's ability to act creatively. This is especially clear as a part of the artistic perspective, where we may see a broad variety of the viewers' reactions to the artist's work. I am referring here to the work being a result of a cultural action—the use of a metaphor. The practice in question is also encountered as a part of the scientific perspective, as I pointed out when referring to research vocabularies. What comes to the game in this case are very specific works—scientific propositions. These scientific vocabularies are rooted in colloquial vocabularies. For example, it is considered obvious that as a part of one of the latter, somebody's travel is justified by its goal. It is most often precise, but no one is expected to identify the methodological grounds behind the construction which called it into being. However, when, say, a biologist appears, these grounds are disclosed. It turns out that it is necessary to refer to the functional relation, which differs from the causal relation; that as a part of this relation, the goal is a necessary condition; that the goal is not anthropomorphic, etc. If now, as a part of the functional explanation,[2] a social psychologist (or a sociologist) analyses Weber's principle of rationality, which provides that individuals pursue their goals undertaking actions—means—that are possibly economical (i.e. based on calculations rather than emotions), they shall not understand the mentality of such persons, when using the starting colloquial vocabulary.

In the case of a colloquial vocabulary, the goal may have a mental character; be understood in the categories of will or desire. Since such vocabularies are transitional, the meaning of the words 'will' and 'desire' is unstable. However, it certainly refers to the human. The goal considered by the biologist does not have such a character. Here, it is the human being who can be the object of study; but it is not their mind, which is reflected on, its domain is not to be used for projecting—vide anthropomorphism—of the mental into the physical. The nature of the goal begins to be understood relationally: as a logical relation between a specific necessary condition and a set of sufficient conditions. Using the biological vocabulary, the sociologist shall continue the relational understanding of the goal and shall avoid anthropomorphism even when pondering on the human mind. The process of the transformation of the understanding of the word 'goal' consists in several steps. The first one is tantamount to breaking the

---

[2] The functional explanation is also referred to as teleological. The term 'teleological' originates from the Greek term for goal—'telos'.

connection between the goal and the colloquially understood human mentality, and to establishing a new relation—between the present event and an anticipated future event. For example, let us ask about the goal behind the emergence of leaves. We shall answer, putting it in a nutshell, that they provide plants with the energy necessary for their further growth. We cannot possibly expect a will or intention from say an oak, but we may assume that—from the logical point of view—its condition in the spring is a sufficient condition behind its autumnal condition.

The second step lies in the establishment of a special domain—the mind—which shall not be described with the help of the colloquial terms mentioned above, but with the help of more technical terms. The word 'intention' does not perhaps have as precise or permanent meaning as the term 'proton', but understood relationally through a reference to the terms 'event', 'propositional attitude' or 'functional relation', shall be more stable than the words 'will' and 'desire'. From now on, the determination of the destination of the journey shall be analysed as someone's intentional identification of an event, which may take place if a mentally—intentionally—acting individual takes further—physical—actions to reach the goal in question. This, however, does not mean that the analysing person is to take into account the human in the events taking place. The human shall constitute a special case of the general explanatory scheme. At the same time, it will be possible to trace the metaphorically mediated transformations of meanings of the relevant terms, which shall be determined by their increasingly new uses. This procedure shall also assume differentiation between metaphorisation in science and the same practice in art, religion or everyday life. The former shall involve verification of the said scientific propositions (sentences) as cultural works. The charm and simultaneously the danger of using metaphors in the other cases lies in the propositions (sentences) present in them either not succumbing to the procedure (religion) or succumbing to it with difficulty (art, everyday life). Let me also recall that the above-discussed falsification is one of the variants of verification, which makes science a domain in which cracks are no lesser painful than in other domains.

7.3 Tension

As already mentioned, culture makes it possible to control the results of cracks, including the cognitive ones, experienced by individuals. However, this is not problem-free. Hence, let me repeat my remark on the individual's resistance against crossing the border. Assuming that only the cultural factor comes into play here—and I am making this simplification consciously—we may assume that people learn to adopt attitudes early enough not to be aware on a daily basis that they determine their non-reflex actions. In connection with the above, they familiarise their environment to such a degree that their sense of safety is not basically problematized. Of course, as suggested above, we may demarcate two extremes—which, by the way, are conventional—one of which is tantamount to complete safety, and the other—its absence. However, I am interested in the multitude of cases between these extremes, where the

7.3 Tension

individual, most often adopting the common perspective, orientates themselves in the world acting in a predictable way, which is only possible when their surroundings are familiarised. Therefore, they can use many tools, including the clocks mentioned above, expect imagined events and solve the encountered problems on the basis of the knowledge they managed to win or acquire. Hence, they act in a stable way. In this sense, I assume that the appearance of an unexpected event constituting a cognitive challenge shall result in resistance against challenging their heretofore existing way of overcoming obstacles. Such individuals shall try to solve the new problem with the help of the old method. Above, I referred to Dewey's figure of the relation between habit and inquiry, assuming after him that this relation is not oppositional but gradable (Zaporowski 2018). Now, I wish to use it to show that the individual in question tends to act habitually.[3]

I am referring to the individual's approach to the problem, which is rooted in well-familiar ways of coping with the problems solved so far. The essence of inductive reasoning, which is used commonly, albeit rarely consciously, inter alia in the common perspective, assumes the projecting of the past into the future. This is so because—be it with reference to many clocks—trust to one's experience is based on long-lasting practice. If, however, one encounters an unforeseeable event, the individual's actions do not move on Dewey's axis towards inquiry, but centre around the habit, which results from the human belief that they are dealing with an exception to the rule. But what will happen if this unforeseeable event begins to repeat itself? In this case, it shall not be enough to disregard the tendency. It will be then that the mentioned resistance shall come into play. The individual shall have to contrast the heretofore existing knowledge—seemingly well-established—to a series of events which do not fit the familiar explanatory scheme. This circumstance will lead the individual, if only they are able to reflect, to ask themselves questions. There will be doubts, but the desire to opt for the familiar will be stronger than that to opt for the unknown. In the meantime, however, the said tendency will continue to exist. The said resistance against correcting one's knowledge will become a sign that the configuration of the attitudes of the reflecting subject will come under tension. The latter will become a function of the conjunction of the intensity of the sequence of unknown events and resistance against ordering them in a way different than the heretofore existing one.[4]

This tension may fluctuate—after all, the said sequence may be, using Rabinow's figure, non-linear, but by persisting it will lead to at least two results. The first one—dogmatization of one's knowledge and disregarding the significance of the encountered, incomprehensible events, will be accompanied by the formulation of ad hoc explanations, so as to minimise the devastating effects of the persisting crisis. Not willing to reject the old order, the individual will want to rationalize the encountered situation in any way. Their habitual attitude will persist, which will make their knowledge fossilized. The second result will have the form of a critique of the existing

---

[3] It was already David Hume who turned attention to this rule.

[4] I am recalling the already mentioned Ohm's law.

knowledge. The tension under discussion will be identified as a mutual incompatibility of attitudes. In other words, a reflecting individual will see that adopting the existing attitudes—with the individual actions testifying to the entire configuration of attitudes rather than any of them separately—results in making mistakes. In the emotional and cognitive sense, the individual will face the consequences of a breakdown of their culture. This culture, being a condition behind subjectivity, will find itself on the brink of breakdown. The ordering of the reality will be disturbed, perhaps suspended. In this case, it is possible—if among the attitudes there is, for example, the attitude of modernity—that the configuration of attitudes will be transformed in such a way as to lower the tension and allow people to order the world of events anew.

7.4 The Rhythmical Tool

The tension mentioned above, being a significant determinant of culture, allows individuals to control sequences (bundles) of events owing to the construction of the ordering tools—a practice being a result of the fluctuation of this tension. I am referring to, inter alia, the said clocks testifying to time. These clocks are marked by the feature consisting in the ordering of events of the physical world which cannot be controlled by people. Nevertheless, they manage to anticipate the appearance of a special sequence (bundle) and in this way, among other things, abstract this particular dimension. Let us take into account a conclusion that time depends on gravitation—for instance 'time passes faster in the mountains than it does at sea level' (Rovelli 2018: 9). Time abstracted from a series of events is not therefore a sole product of human thought.[5] At the same time—regardless of gravitation—it is possible to record the rhythm in which time is measured. I have mentioned above the experience of Conrad, who became aware that the rhythm of the drum replicates the rhythm of the heart. Let me add that more or less three hundred years earlier, Galileo came into contact with a certain series of events. While being in the cathedral in Pisa, he saw a lantern, which was moving. He simultaneously started to measure his pulse and noticed a similarity between both these rhythms (Gregersen 2021). In this way, he constructed a pendulum. The heart—unless someone suffers from arrhythmia or some other condition—is marked by a rhythm, which may accelerate or slow down, but is regular.

The pendulum and the drum acquire the same property—isochronism—but it is in particular in the case of the latter that people can easier control the acceleration and slowing of the rhythm. Let us, however, notice, that the work of the heart consists in the alternate contraction and relaxation of the muscle. This process can also be observed in the work of the pendulum, which moves alternately in two directions. In the case of the drum, the system is simplified, since motion is based on the repetition

---

[5] Hans Reichenbach stated that '(...) our knowledge of time is not a priori, but the result of observation' (Reichenbach 1951: 155).

of the sound followed by a constant interval. After all, the sound of the instrument replicates the pulse rather than the actual work of the heart. The latter is repeated in the work of the pump, a steam engine or an internal combustion engine. We know that the heart is a functional equivalent of tools generating the movement of liquids and gases. Anyway, even the checking of the pulse allows us to abstract time owing to the repetition of the same cycle. I therefore believe that one of the simplest and at the same time most effective ways of ordering physical events is approaching them through a recreation of rhythm. This ordering does not simply lie in the recording of the particular subsequent events so as to in turn generalise the sequence. This work is more of a determination of the relation between two opposite movements—as in the case of the heart—or between movement and its absence in a specific cyclical order. Only the determination of this relation allows one to anticipate a subsequent event.

The rhythmical tool does not therefore have to be designed to recreate time, but replicating it, it can also order special events. I am referring here to the ones initiated by the human being when producing new worlds, including the social world. I have presented one of such tools above—it is that of custom, which is closely related to the clock showing cyclical time. On the one hand, custom imitates such a clock, just like, for example, the beaten drum imitates the beating heart. There are customs overlapping the solar calendar. For instance, Christmas takes place around the winter solstice (in the Northern Hemisphere), while the Slavic Kupala Night— exactly during the summer solstice (also the Northern Hemisphere). On the other hand, the use of customs is integrative. This is because it consists in a cyclical gathering of the community—this action assumes the identification of a worldview-related goal. Such a goal is assumed by the already mentioned customs, but for example the Celtic Samhain or the Slavic *Dziady* (Forefathers' Eve) celebrated on the night from 31 October to 1 November—both being festivals devoted to the dead— assume the end of summer understood in an agrarian rather than astronomical sense. The former testifies to a special form of social organisation rather than the recording of heavenly bodies. Anyway, the custom sustains social cohesion owing to the rhythmic repetitiveness of the same pattern.

7.5 The Change and the Crack One More Time

Custom is intertwined with myth. Let me repeat that it is designed to blur the non-continuity of the world (or experience)—this takes place due to the closed character of narration. If, therefore, we put the two forms together, it will turn out that the replication of the necessary nature of the sequence of events makes this complex system a very effective tool consolidating the group. Let us refer to the above-mentioned Christmas. The festival can be understood as a bundle of cyclically replicated events: the Christmas Eve supper, singing carols, the Midnight Mass, etc. The individual co-creates these events if they become skilled in the appropriate behaviours and recognition whether others perform them in the same way or not. In this sense, custom-related actions are routinised to the point of being performed unconsciously, automatically. At the same time, the festival assumes a certain world-view-related message. Participants of the Christmas Eve supper and the Midnight Mass refer to the certain content—the myth on the birth of Messiah—and the content makes the custom sacral. Also in this case, they act in a learnt way. Nevertheless, the very myth does not have an action-specific nature; it is a cultural work. In this sense, probably every custom assumes the presence of a myth, although not every myth necessarily assumes the presence of the custom. At the same time, the parable in question belongs to the ones, which—to repeat myself—are not only closed, but also circular. This makes it possible to entwine myth not into just one, but a series of customs located in various points of the annual cycle,[6] which further strengthens the integrating power of this special custom-related/mythical system.

Both myth and custom are subject to transformations. For instance, Halloween combines the clearly separated heritage of pre-Christian (including the already mentioned Samhain), Christian (by preceding All Saints' Day) and post-Christian customs. In this last case, we are talking about, inter alia, the feast acquiring a commercial character. The transformations experienced by Halloween assume the emergence of changes, including cracks, while Christmas seems to be resistant to the latter, i.e. is marked by a relative cohesion. Every cycle is more or less regular. This means that the process of the replication of the circular movement lies in referring to events covering the relation between two states which are the same as before. With reference to my previous remarks, two processes can be presented. On the one hand, the work of the heart assumes the change—contraction as the first one and relaxation as the second one—which together make up a feedback process. On the other hand, the drum and custom assume change—a relation between the sound and the interval, and between the feast and the time after it; this process is not based on feedback and is cyclical—it lies in a repetition of a distinct scheme. However, the process can be disturbed. A crack is tantamount to the appearance of a new scheme, and it can be visible to a larger or lesser extent. What differs Halloween from Christmas is that the former, having a mass character, is not 'sustained' by various institutions (organisations) such as Churches, which would 'legitimise' its form and in this way would also create an illusory sense of a continuity of a given custom. Anyway, the crack is a change which makes it impossible to replicate the process in its heretofore existing form. From then on, the tool will no longer work or will work according to a different key. Significantly, the latter option may be tantamount to another cyclical work. Indeed, this is testified to by both feasts mentioned above.

The crack is therefore of primary importance for the process of the ordering of a stream of events, in particular when the problem of non-continuity of the world is referred to. Let me recall that Kołakowski analysed myth as a means used for the maintenance or a restoration of this continuity. I also wrote about scientific concepts some of which blur, and others emphasize the dimension of time. If, therefore, the

---

[6] It is commonly known that some elements of such a series, like Christmas, are fixed feasts, while other, like Easter—movable feasts. However, the relation between the former and the latter has a regular nature.

crack constitutes a possible (and frequent) rather than necessary element of the experience of the world of events, then a number of scenarios of coping with the change appear. Firstly, a new myth emerges in place of or next to the old one – let us consider the already mentioned narrations on the Messiah of Jews and Christians—which, wrought into customs, maintains its cyclical character and is as much dogmatic as therapeutic. Secondly, a scientific concept (theory) is constructed—for instance, one authored by de Boulainvilliers or Lévi-Strauss—which also has a cyclical nature, but takes on a critical quality. In both cases, time is disregarded, while both the myth and the concept seem to assume the continuity of the world of events. Thirdly, yet another researcher—may it be Frazer or Rabinow—creates a theory, which breaks the cyclical nature of narration, while possessing a critical quality, with time highlighted as a plan on which the protagonist of the tale moves in a given direction, not necessarily linearly. The world of events ceases to be based on continuity, and the future becomes a projection of the past or something open. Can a different scenario be imagined?

7.6 Two Elements and Direction

Let us begin with the passage of time, which seems to be sensed intuitively. Remembering about the excessive non-clarity of colloquial terms, let me immediately make it clear that I am referring to an irreversible change in time. This issue is testified to by the figure of the crack, as well as the figure of the border. As a part of two scenarios, the change is present, but its irreversibility is apparent in the sense that a unidirectional sequence of one state after another is ceaselessly replicated. I have already mentioned the disregarding of time, because the tool referring to the figure of the rhythm or the balance gives the sense of its absence. In reality, this is about the sense of the absence of irreversibility of the said change. This quality is in turn underlined as a part of the third scenario, but in this case, it is the cyclical nature of experience which is disregarded. In other words, whether it is the unidirectional movement with a purpose, or with no purpose at all, each of the three stages is unique, which allows us to suppose that it is contingent. What is therefore disregarded as a part of this scenario is the order of events, or, to be more precise, the sequence of one event after another. The latter successfully records the tool which assumes a cyclical appearance of a testimony of a given assumption. I would therefore like to propose a scenario, as a part of which the observer—oscillating between different perspectives, but potentially ready to adopt the scientific perspective—orders the world of events with the help of the figure of the cycle, while perceiving the fluctuation of this order with the help of the figure of the crack.

Let us imagine a group of individuals rather than a single person, who, while interacting with each other, verify the use of propositions—cultural works—of the partner in a given contact, which testify to the identification of physical events.[7] Such

---

[7] I treat this as a special case of triangulation as interpreted by Davidson (2001b).

a network of interactions is to ceaselessly check whether, for instance, one's own research into the current is not illusory. After all, we are talking about the process of a possibly continuous recording of events—it would be hard to reflect on it ceaselessly. And this is, among other things, what the problem of maintenance of the balance between a part and a whole is about. May then several individuals temporarily agree that a specific fragment of the world of events is ordered in such a way as to provide, inter alia, the predictability of one event taking place after another in a cyclical way. May this group use a tool allowing identification of this fragment. In contrast to the above-mentioned works, the tool shall allow expression of generalised propositions on the identified order. May this group consist, for instance, of farmers cultivating olive trees. Let us assume that their knowledge, consisting of propositions mediated by their predecessors, as well as propositions resulting from their own observations, provides that the trees must be regularly supplied with water so as to bear the possibly most handsome fruit each autumn.[8] In potentially unfavourable conditions, they will cyclically, year after year, use a system of artificial irrigation keeping the plants in possibly good condition.

In the meantime, let us imagine that this order has broken down. Olive trees are farmed inter alia to acquire olive oil. For some time now it is believed that olive oil is therapeutic owing to its components called polyphenols,[9] and that at least in the case of a certain Portuguese variety of olive tree, irrigation lowers their levels (Machado et al. 2013). This conclusion may result in a crack and the adoption of the border attitude. Hence, I assume that some—or perhaps all—members of the community would have to either consider such a break an illusion or an erroneous understanding of events, or cope with the necessity to harness the tension between the heretofore existing attitudes, and generate a new order. In the latter case, I postulate that these people—perhaps creating other temporary groups or individually—would use a tool, which, as before, uses the figure of the cycle for the identification of relations between the particular events, but assumes a correction of the earlier course. In the case under analysis, the correction would lie in relying on (natural) rainfall as a source of water. Let us consider another example. The terms 'sunrise' and 'sunset' are still commonly used to indicate the cyclical nature of the movement of the sun. At the same time, we know that this movement is illusory. Let us reflect on the fact that 500 years ago, Copernicus did not reject the figure of the cycle; he changed its form, while making it more precise. Hence, I assume that it is not the figure itself, but its use, which is a problem. Simultaneously it turns out that our predecessors created an ordering tool with a much more far-fetching character than they might have thought.

---

[8] Olive trees are grown in the Northern and Southern Hemispheres, so when writing about the autumn, I am referring to the former.

[9] For example, the European Commission issued a Commission Regulation No. 432/2012 referring to this issue.

7.7 With No Retreat

May the increasingly new orders be in place between the subsequent crises, and at the same time, may many tools—cultural works—covering various parts of the world and mutually overlapping, appear. May then this situation refer to different perspectives and people—the latter can be the creators of the tools used in some perspectives, and their users in others. In this way, a network of alternative, but also mutually complementing ways of grasping the relations between events will emerge. In this entire network of cognitive results and acts, a special place is occupied by the aforementioned crack. This is because it will be a witness to crisis. However, what will be the consequences of the coexistence of systems which are in conflict with each other in the context of the occurrence of the event under discussion? Experience shows that on the one hand, advanced scientific propositions can be concurrent with archaic commonplace propositions—vide the vocabulary containing the term 'sunrise'— which testifies to a relative tolerance to putting together cultural works, which, treated separately, are handy and effective. On the other hand, putting them together does not make each of these works handy and effective in all circumstances. In this sense, as much as their use overlaps as a part of different perspectives, then if someone experienced a crack in relation to any of them in any area, then its use in this particular area ends.[10] This end is grasped *ex post* and is related either to the sense of a loss, or an ability to order the reality in a new way, although we cannot exclude a coexistence of these two variants, either. Anyway, *stultitia* and falsification are consequences of the crack rather than something which precedes or accompanies it.

If then a possibility to order the world in a new way emerges, it will turn out that the crack assumes an irreversible movement, but its direction is not yet certain when the crack is taking place. I am not referring here to the issue of Rabinow's non-linear space, but I think that the demarcation of any direction is very risky under such circumstances. Let us imagine that we are researchers having a hypothesis on a section of the world. The hypothesis, together with the presumed rest of hypotheses, allows us to explain and anticipate events. If we wanted to extend our knowledge, we would have to construct a better tool. For instance, let us refer to Popper's position, which was mentioned above (Popper 1979). It provides that the impossibility of classification of a more unlikely hypothesis, which is simultaneously richer in content than the one we have, resembles the effect of stepping beyond the Foucauldian border. However, the relation of correspondence establishing that the old hypothesis proposes making what the new hypothesis proffers as more familiar, assumes the presence of one border rather than a sequence of them. It is similar in the case of other methodological concepts. Obviously, when reading Foucault, one may assume the crossing of subsequent—implicitly many—borders, but using a tool, it is difficult to anticipate a series of more effective or successful uses, if not the end of the use. Also, Kmita's writings show that the author testifies to one—fundamental—crack:

---

[10] In this context, I believe that at least as a part of the scientific perspective assuming a critical distance to oneself, the experience of the end of the use of a tool is not devastating to the user. However, it is certainly so from the religious perspective.

the parting of the symbolic and the technological spheres. The question about the future remains open. I believe that we are dealing here with a reference to Weber, who, when writing about breaking the spell put on the world—the end of magic—referred to a single crisis rather than a series of them, while being conscious of the approaching crisis of rationality, with the consequences he could not possibly know. It was only Habermas who turned attention to the latter.

The problem with the demarcation of the possible direction of movement, as testified to by the crack, lies in the necessity to take into account at least two such changes. A single crossing of the border is tantamount to no more than a departure—certainly a methodical rather than an accidental one—from the heretofore existing practice, but does not allow the establishment of any stretch that would determine at least a temporary azimuth. Nevertheless, the crack assumes a move of leaving behind one's previous way of orientation in the world, which already allows one to make comparisons and possibly plan a subsequent step. Therefore, not opting for moving even in the said non-linear space, I postulate that the crack testifies to the move, which has irreversible consequences. Crossing the cognitive border does not resemble an action accompanied by counteraction, as in the case of the beating heart. The latter is a device which owing to the alternate occurrence of oppositional changes makes the system stable. The modern individual uses this stability to apply it to new situations and thus broaden their cognitive field. It is similar in the case of tools such as the sundial, which despite not being based on a system of opposite actions, are also based on the mechanism of the cycle. Therefore, it turns out that this mechanism is universal enough to be useful after the cognitive crises. Whilst the crack is irreversible, the principle of the cycle is not.

7.8 Reconstruction of Events

The discovery of the principle of the cycle in nature does not mean that showing it with the help of the figure of the circle is accurate. This practice has an idealising nature. After all, the spiral or ellipse-like movement shows the behaviour of bodies in the micro- and macro-scale in a better way. At the same time, although the figure of the circle is a good representation of a turn or pulse, it certainly does it approximately. This is so because people project the effect of abstracting properties from the experienced events back into the world. The circle as a geometrical figure is a cultural work designed, just like propositions (including scientific ones), to order the reality, which potentially slips out of their control. However, we are talking about the figure, which has some very practical uses. It is common knowledge that movement which is not maintained, eventually comes to an end. This applies to, inter alia, human life, which passes away. But not just human life. We may use as further examples the figure of the Milky Way, a water whirl (or whirlwind) or the snail's shell. The shape of the spiral is not accidental in the sense that its subsequent curvature, lying increasingly further from the starting (central) point, testifies to, respectively, the stage of the shell-forming process increasingly further from the current stage and the decreasing speed of the bodies co-creating the arm of the galaxy. The whirling movement makes it possible to notice how the external masses of water and air are pulled into the inside, down the curvature with an increasing speed. The above-mentioned form of the hermeneutic circle is read in the opposite direction. This is because the unwinding of the spiral is, in this case, to indicate the development (extension) rather than regression (tightening). May also this comparison be a meaningful signal that a given cultural work—in this case the figure of the spiral—should always be referred to its use. Anyway, the figure of the circle does not recreate the actual processes. As mentioned above, this is done better by the figure of the spiral, which pictures the slowing down of the movement—a gradual cooling of the body.[11] It is not therefore accidental that the figure of the circle turns out to be an attractive tool, which assumes both change and rhythm.

The above situation suggests that it is possible to blur the end as a real condition of the world of events, including the ultimate destiny of the human being. In this context, I wish to indicate two aspects of the said attractiveness of this ordering tool. Firstly, taking into account the world possibly close to sensory experience—the transformation of the day into the night, sunrises and sunsets, the pulsation of the beating heart—the figure is appropriate to reflect the change taking place. After all, it was already the Babylonians who divided the annual cycle into twelve parts, which we also use, so as to successfully order the experience. The order of the phases of the Moon is no less archaic. If today we know that the 'croissant' we can see from time to time on the night sky is only an impression—a reflection of the sunrays from a part of the surface, the rest of which is hidden by the curvature of the Earth—rather than a real celestial body, this does not change the fact that the effect appears cyclically at the same time intervals; that it is regular. Similarly, a doctor or a paramedic, who, while checking the pulse, establishes rather than supposes that a motionless individual is alive, refers today to the same experience which allowed people distant in time to use the instruments replicating the rhythm of nature and feel its integral part. Because in the horizon of the events taking place possibly close to the individual, a deviation from the recreated rhythm (scheme) is unnoticeable, it is understandable that the manner of the recreation, which turns out to be effective, is considered certain. Let me therefore repeat that despite new data indicating that such calculations are highly approximate, they continue to be practiced.

Secondly, this calculation has yet another use. People can notice not only the cyclically appearing full phases of the Moon, but also the said passage of time. Therefore, they can use a specific tool not only to reconstruct the cycle, but also to manipulate it. In this sense, time does not have to measured; it can be forgotten. As much as the first use of the tool can be called technological, the last one can be referred to as—and I have already used the term—therapeutic. Most often, however, we have in mind the word 'symbolic', as testified to by Kmita's example. Anyway, this is about the creation of a tool, which will make the change as much replicating as sustaining the vitality of a given body, including the social one. Myth captured from the point of view of custom is integrative, since the people who share it undertake

---

[11] Principles of entropy can be a good example here.

joint actions on a cyclical basis. They do not necessarily notice that the tool they use recreates a rhythm. However, acting together, they order, inter alia, other dimensions of their life more effectively and in further-reaching terms than if they did it single-handedly. In this sense, the therapeutic (symbolic), including custom-related nature of myth has its consequences for the development of technology—at least one based on the methodical use of regularity. On the other hand, a question appears whether the noticing of the replicating rhythm of nature does not allow one to project it, via customs clothed as myth, into the life of the individual and the continuity of the group, thus giving the life and the continuity of the order the form of a cycle rather than a figure reflecting the passage of time. In this sense, I assume inter alia a feedback between technology and myth.

7.9 Freedom

What therefore happens on the individual or group horizon, is the experience of, inter alia, the cycle. I am using the term 'inter alia', as I am far from claiming that this is the only experience.[12] However, I believe that it is fundamental for the constitution of the identity[13] of both the individual and the group. This experience is ordered owing to the use of a very special tool—the circle. The individual or the group—or individuals as social animals—perceive themselves and the world around them as a cyclically replicating system. However, the experience also covers a crack of the rhythm. The cycle is broken. On the one hand, we may be dealing with an accidental event (or a series of events). It can simultaneously be traumatic, which results in a creative, including an artistic, testimony, as exemplified above. However, on the other hand, this experience can be intentional—in particular when the individual or the group such as one acting in a scientific perspective, perceives that the work of the device replicating the rhythm is no longer effective. The practice of 'social engineers', who are well-grounded in the religious perspective and who create new tools to consolidate or even create communities to perform specific tasks is yet another option. May this be backed up by Brubaker's remarks on the organisations constructing groups using the relation of war as a tool (Brubaker 2004). In this case, it is about a crack of the cycle of the persistence of social solidarity at the local community level and replacing it with a cycle of the rule of social solidarity at the level of ethnicity or nationhood.

Anyway, the experience of the cycle begins to be accompanied by the experience of its disturbance. Reactions may include resistance to changing the heretofore existing practice or a change consisting in the demarcation of a new trajectory of one's

---

[12] The figure of the cycle assumes the ordering of the experience of regularity in a particular rhythm (in time). Let me recall, however, that it is possible to think about regularity, disregarding time. Examples may include the above-mentioned custom of shaking hands and Quine's observation categorical.

[13] I understand identity in a classical way, in two terms: ontological and logical. The former is about the consequences of Parmenides' statement 'A is', and the latter about the consequences of the use of the formula 'A $\rightarrow$ A'.

actions. Also, a possibility emerges—including a tendency of some anthropologists as described above—to replace the figure of the circle with the figure of the road, assuming at least three steps: in the linear or non-linear space. What is characteristic is that walking down such a road is tantamount to not returning to the stage already left behind. It is similar in the case of the scenario as a part of which a new cycle constitutes itself after a crack. What differs a proposal of a new cycle and a proposal to begin a new stage of the road is that the events assumed by the former are—naturally hypothetically—ordered. The latter proposal makes ordering an open question. At the same time, nothing prevents the scenario assuming a replication of a new cycle of events preceded by a crack of the old scheme from not testifying to the road covered so far. However, such a road can be presented—as mentioned earlier—*ex post*, step by step rather than as a ready model based on the tripartite relation of succession. Above all, however, I consider the circumstance under which the experienced crack itself can become a tool, just like the experienced cycle became a tool earlier, as significant. I am referring to a conclusion of at least some people that disturbance of the heretofore existing order can be managed. From then on, the crack can be used—in a variety of ways.

Kmita indicated circumstances of the parting of two systems of cultural works, while Foucault indicated the circumstances of undertaking a single cultural action. These reflections complement each other, enabling the creation of a picture of the human being—an inter alia social animal—who not so much manipulates the crack, as is sometimes its object, and its subject at other times. I postulate that this picture assumes the presence of culture understood as a configuration of attitudes, including propositional ones, determining non-reflex actions. An action is cultural when its components include the adoption of an attitude—a configuration of attitudes—with a bundle of such actions resulting in the construction of cultural works. The latter are effective only when someone can use them, even if they did not create them. It is also in this case that a reference to culture is indispensable, because its use is tantamount to the undertaking of a relevant cultural action. In this sense, the construction of the tool and the use of the tool are two terms for the same process. If, on the basis of experience and calculation based on the ability to abstract, the human being managed to create such tools as the cycle and a disturbance of the cycle, and use them, then culture turns out to be a system which is vulnerable in the sense that it is exposed to manipulations. If, however, people come to the play—by which I mean modern people differentiating between the symbol and the technology and at the same time referring them to one another—who are conscious of the consequences of the use of cultural tools, then at least it can be assumed that their freedom is not tantamount to the necessity to manipulate others or being manipulated by others.

References

Brubaker R (2004) Ethnicity without groups. Harvard University Press, Cambridge, London
Davidson D (1978) What metaphors mean. Crit Inq 5(1):31–47

Davidson D (2001) The second person. In: Davidson D (ed) Subjective, intersubjective, objective. Oxford University Press, Oxford, pp 107–121

Gregersen E (2021). http://www.britannica.com/technology/pendulum. Accessed 18 Oct 2021

Machado M et al (2013) Polyphenolic compounds, antioxidant activity and L-phenylalanine ammonia-lyase activity during ripening of olive cv. "Cobrançosa" under different irrigation regimes. Food Res Int 51(1): 412–421

Popper KR (1979) Conjectural knowledge: my solution of the problem of induction. Popper KR objective knowledge. Oxford University Press, Oxford, pp 1–30

Reichenbach H (1951) The rise of scientific philosophy. University of California Press, Berkeley, Los Angeles

Rovelli C (2018) The order of time (trans Segre E, Carnell S). Riverhead Books, New York

Wittgenstein L (2001) [1922] Tractatus logico-philosophicus (trans Pears DF, McGuinness BF). Routledge, London, New York

Zaporowski A (2018) Action, belief, and community (trans Moroz-Darska A). Peter Lang GmbH, Berlin

Chapter 8
Conclusion

Abstract In conclusion the author stresses the inevitability of the passage of time. While presenting the competing representations of such a passage the figure of the cycle is preferable. However, one must not forget that the cycle may not only accidentally be broken due to various cracks. It is then possible that one's system of attitudes—culture—may allow one to arrange a new cycle to take new directions along one's journey.

Keywords Time · Cycle · Crack · Culture · Attitude

Above, I have presented various scenarios of coping with a representation of movement to which, inter alia, people are subjected. I am referring here to the picture of the inevitability of the passage of time. I have postulated that these scenarios use the figure of the cycle or the road. The former variant means that movement and, further, time, is disregarded. Interestingly, tools using the figure of the closed cycle are encountered in the religious and scientific perspectives. In the former case, this is about myth, which often blurs the experience of the passage of time through custom. The same applies to the scientific position, the author of which assumes the ordering of events with the help of a theoretical model, distancing themselves from taking into account the temporal dimension in the context of the abstraction of notions from the empirical testimony. The basic difference between these cases lies in, respectively, the therapeutic (symbolic) and the critical (technological) nature of the constructed and used tool. Tools referring to the subsequent figure also have this latter character. In this case, its constructors adopting, as before, the scientific perspective, do not disregard time. Just the opposite—they highlight its role, although two variants appear here, as well. The former assumes that the road we travel has an end. Hence, the figure used has a both necessary and developmental character. The second variant refuses an end to the road, which thus becomes contingent, being subject to mutations. I assume that neither of these figures is used solely to allow their constructors and users to cope with movement as such. What comes into play is a disturbance of this movement.

© The Author(s), under exclusive license to Springer Nature Switzerland AG 2025
A. Zaporowski, *Crack and Culture: On Representations of Movement in Anthropology and Philosophy*, SpringerBriefs in Philosophy,
https://doi.org/10.1007/978-3-031-83422-6_8

This is why I have referred to the crack as an event—a testimony to an irreversible change, which makes the passing of time an even more painful problem. A number of processes, including human life, turn out not to be a permanent cycle or a permanent road. The mythical and a priori tools on the one hand, and the empirical tools on the other, indicate different concepts resulting from the experience of this event, with its traumatic nature being underlined by artistic works. My position is that I assume, firstly, that the crack is a testimony to not so much suffering as inaccuracy of the heretofore existing cognitive practice. In this sense, I treat them as an indispensable element of a critical attitude to this practice, as shown by, inter alia, Foucault's attitude of modernity. Secondly, the crack does not have to be accidental; it can be generated to check to what extent the current tool is technologically effective. Here, it is worth referring to Kmita, who showed that this check must assume a relative independence of the technological sphere from the symbolic sphere. My position is one of making an attempt at a simultaneous highlighting of the role of time and presenting it with the help of the figure of the cycle. I thus recognize that the cycle does not have to be used to blur time. I do not think that it should be necessary to refer to the mechanism of a replication of a given state or process. On the other hand, I do not think that time is to be presented linearly or non-linearly out of necessity, always in the form of a curve, the ends of which do not meet each other.

Of course, intuition tells that it is the linear presentation of time which is appropriate. This is because the phrase 'passage of time' was not coined accidentally to indicate movement, direction and the impossibility of return. However, a more critical stance leads to a conclusion that the forward movement and change can be accompanied by a cycle. And if this is so, then movement is understood two-dimensionally. The body becomes an element of repeating events and new events. In the former case—as testified to by changes such as a turn of the Earth or lunar phases—the body remains in the state of balance. In the latter case—in turn, as testified to by the crack—the body is forced to adopt a new direction, looking for a new state of balance. A unidirectional interpretation of movement in the categories of the road makes it impossible to grasp this condition clearly. This is because a given stage is ordered (or not) because of another stage rather than the clock, which testifies to its (be it very brief) existence. For example, a given hypothesis is maintained not so much because of its regular confrontation with events, as in connection with its confirmation or falsification by accidental events. In the scenario I am proposing, a cyclical confirmation does not exclude a sudden falsification. However, I can think of a situation in which a periodical stability is ensured; its possibility is assured by the clocks mentioned by Einstein. The problem lies in how to broaden the knowledge, being aware of a limited usability of these clocks.

This limited usability does not however apply to the cycle as a mechanism. This is testified to by the already mentioned example of Copernicus. If, in turn, we assume that the cycle does not last indefinitely, but is subject to cracking, we shall understand that we may go ahead, experiencing these cracks and changing the subsequent states of balance. However, as indicated above, the delineation of such a path is only possible *ex post*. It is also a result of a crack of a series of experiences and calculations. On the one hand, a number of individuals notice not only movement itself, but also its

8 Conclusion

generation and withering. In this sense, they repeat, one after another, the phrase on the passage of time. On the other hand, time, which turns out to be a property of the body which recreates itself cyclically, is abstracted on the basis of a number of clocks. In this sense, the significance of the construction of the pendulum is understandable. A perfect replication is, however, an illusion, since the body slowly becomes motionless. For this reason, what emerges is a desire rather than a necessity to sustain the cycle. This attitude may, although it does not have to, carry with it a number of advantages. I am not referring to, for example, a group-consolidating nature of the worldview. Let us notice that when funding a community, many myths ruin other communities. What I mean is the use of the said figure of the crack, which is a warning sign of the temporary nature of a given cycle. It allows to maintain a critical distance to practice, including the scientific one. What I have in mind, inter alia, is an examination of the current, which should be accompanied by an anticipation of something which has not yet taken place, but when taking place, will be different from the already familiar.

Culture is of a fundamental significance in this context. It is a result of the human being's interaction with the world of events. In this sense, it is necessarily learnt. Culture makes it possible to order the said events, although I understand the relation between them bidirectionally. Experiencing events, people gradually begin to maintain balance—the more so the more they are exposed to their pressure. However, not everyone experiences the same bundles of events, especially that some of them are constituted by non-animate bodies, while others are animate, including social ones, with the range of social structures being broad. On the one hand, it is possible to imagine an individual who is an opportunist reacting to stimuli in a routine way. On the other hand, there appears an individual creative enough to use their past experience for a relatively constant rebuilding of their system of attitudes so as to themselves become a challenge to these stimuli. In this sense, culture can be interpreted as a possibly flexible work. In connection with the above, I treat a crack as a special warning signal and a call for breaking with the encountered systems of attitudes, which seem to reflect the experienced order. For example, it is possible to derive a socially sanctioned agrarian cycle from at least some bundles of events founding the solar cycle. Culture enables a separation of the figure of the cycle itself from the content that is attributed to it and its use for the construction of new figures, including the figure of a cracked cycle, which allows at least critical individuals to take new directions along their journey.

This journey can be presented in a variety of ways. At the beginning of my reflections, I proposed an image of the river. It was a highly personal tool, but, I believe, one used not very rarely to reflect not only movement itself, but also the passing of time. In the context of the analysed positions, in particular ones which, when placed in the scientific perspective, result themselves in tools allowing to present the issue of my interest, as well as my proposal, I wish to reflect that the presentation of my life as a journey down a river may become transformed. First, when moving forward, I can only guess the further course; I consider the metaphorisation woven into imagination as a way of not only reacting to but also evoking future events, which will constitute this course. Secondly, it seems to me that the heretofore journey is a series of cycles of experience of the changing surroundings. What I have in mind is a situation as a part of which, while moving forward, I periodically act on the basis of the same system of attitudes, to act differently, changing this system, after initially usually accidental and subsequently often planned cracks. I do not think that each crack results in a better, for instance more effective, way of coping with the encountered events. However, I do believe that making mistakes and experimenting does more good than passive surrendering oneself to these events, including other people's actions. Thirdly and finally, while moving forward, and at the same time cyclically, I discover that the said missing film frames are a result of the inability to harmonize two movements simultaneously. There is still a lot for me to learn, but I am streaming forward.

Index

A
Adler, Karen, 59
Ajdukiewicz, Kazimierz, 6, 41
Anthropology, 10, 23, 34, 36, 37, 40, 49, 50, 56, 68
Attitudes, 2, 3, 5, 6, 9–12, 17, 19, 24, 29, 32, 33, 35, 39, 48, 51–54, 57, 61, 62, 64–69, 71, 73–81, 84, 85, 87–92, 96, 101, 104–106

B
Bielawski, Jozef, 24
Border, 7, 11, 12, 32, 53, 71, 81–85, 87, 88, 90, 95–98
Brubaker, Rogers, 65

C
Cassirer, Ernst, 18
Change, 3, 5, 7, 24, 30, 35–37, 39, 41–50, 52–56, 58, 60, 63–65, 68, 71–79, 82–85, 87, 88, 94, 95, 98–100, 104
Circular, 8, 18, 28, 45, 46, 58, 94
Clifford, James, 68
Cohen, Leonard, 58
Conrad, Joseph, 7
Crack, 3, 4, 7, 54, 55, 57, 59, 60, 64, 65, 71, 81, 82, 87, 88, 90, 94–98, 100, 101, 104–106
Cultural action, 76, 77, 79, 81, 84, 85, 88, 89, 101
Culture, 2–4, 6, 7, 12, 13, 19, 23, 30, 36, 47, 53, 61, 62, 65, 69, 71–82, 84, 85, 87–90, 92, 101, 105

Cycle, 3, 8, 25, 26, 28, 44, 45, 55–57, 83, 87, 93–96, 98–101, 103–106

D
Davidson, Donald, 13, 68
Durkheim, Émile, 16
Dylan, Bob, 84

E
Event, 3, 4, 6, 8–11, 13, 23, 26, 27, 33, 34, 39, 43–45, 47, 48, 50–52, 54, 56, 59–65, 67–69, 71, 73–77, 81, 84, 85, 88, 90–101, 103–106
Explanation, 14–16, 28, 52, 63, 65, 89, 91

F
Foucault, Michel, 5
Frazer, James, 26

G
Geertz, Clifford, 9, 13, 20, 39, 46, 65, 71, 81
Gellner, Ernest, 6
Gregersen, Erik, 92
Gudykunst, William, 60

H
Hall, Edward, 9
Hume, David, 14

I
Interpretation, 15, 17, 19, 25, 43, 53, 56, 72, 85, 104

J
Joyce, James, 41

K
Kim, Young, 61
Kmita, Jerzy, 6
Knowledge, 5, 6, 10, 19, 21, 24, 26, 27, 34, 44, 59, 61, 62, 64, 65, 68, 69, 71–73, 88, 91, 92, 96–98, 104
Kolakowski, Leszek, 8

L
Lévi-Strauss, Claude, 11
Linear, 8, 10, 11, 18, 23–25, 28, 31, 33, 40, 45, 58, 87, 91, 97, 98, 101, 104

M
Machado, Manuela, 96
Malinowski, Bronislaw, 75
Mandelstam, Osip, 55
Marcus, George, 68, 80
Mascarenhas, Fernando, 68
McCormack, Mike, 80
Metaphor, 1, 20, 88–90
Movement, 1, 5, 8–11, 14, 23–28, 30, 33, 36, 37, 39, 45, 46, 49, 51, 56, 78, 84, 87, 93–99, 103–106

N
Nabokov, Vladimir, 57
Nagel, Ernest, 14

O
Oberg, Kalervo, 60

P
Popper, Carl, 14
Propp, Vladimir, 11
Proust, Marcel, 56
Putnam, Hilary, 47

Q
Quine, Willard Van Orman, 12

R
Rabinow, Paul, 31
Reddy, Deepa, 34
Rees, Tobias, 36
Reichenbach, Hans, 92
Representation, 1–5, 8, 11–15, 18, 27, 39, 41, 45, 47, 62, 63, 66, 98, 103
Rhythm, 3, 5, 7, 8, 10, 11, 43, 44, 58, 66, 92, 93, 95, 99, 100
Robinson, George, 59
Rovelli, Carlo, 92
Ruben, Brent, 61

S
Stavrianakis, Anthony, 31

T
Toker, Leona, 57
Twardowski, Kazimierz, 6
Tylor, Edward, 23, 71

U
Understanding, 12, 16–18, 21, 30, 32, 33, 35, 37, 50, 56, 61, 72–76, 89, 96

V
Verification, 27, 46, 63, 87, 90

W
Weber, Max, 6
Wittgenstein, Ludwig, 50, 72

Z
Zaporowski, Andrzej, 15, 26, 27, 29, 31, 32, 45, 47, 50, 56, 69, 73–75, 83, 91

www.ingramcontent.com/pod-product-compliance
Lightning Source LLC
Chambersburg PA
CBHW050020170225
22051CB00003B/30